Role of Cell Surface in Development

Volume II

Author

K. Vasudeva Rao, Ph.D.
Professor
Department of Zoology
University of Delhi
Delhi, India

CRC Press, Inc.
Boca Raton, Florida

Library of Congress Cataloging-in-Publication Data

Vasudeva Rao, K., 1933—
  Role of cell surface in development.

  Bibliography: p.
  Includes index.
  1. Cell membranes. 2. Developmental cytology.
I. Title.
QH601.V385   1987   574.3   86-13684
ISBN-0-8493-4687-8 (set)
ISBN-0-8493-4688-6 (v. 1)
ISBN-0-8493-4689-4 (v. 2)

International Standard Book Number 0-8493-4687-8 (set)
International Standard Book Number 0-8493-4688-6 (Volume I)
International Standard Book Number 0-8493-4689-4 (Volume II)

PREFACE

Developmental biology has now emerged as a truly interdisciplinary science. Classical embryology of past centuries has revealed a wealth of knowledge on how animal embryos develop. From this knowledge comparative embryologists have reached such far-reaching generalizations as the homology of germ layers. Research in embryology during the present century is characterized by the experimental approach. Interfering with developing embryos in various ways has been the essential feature of the work of experimental embryologists. In the past two decades many new techniques developed by biochemists and biophysicists have contributed greatly to understanding the mechanisms of development. A developmental biologist can now expect to find a rational explanation for embryological observations which once looked mysterious. Progress in a multidisciplinary science needs to be surveyed periodically, not only to take stock of the achievements, but also to indicate profitable future approaches. A restatement of old problems often brings into sharper focus the area where rapid progress is possible with the newer techniques.

Recent progress in the area of membrane biology amply justifies a discussion of animal development to highlight the role of the cell surface. I have had to survey a fairly wide area of modern biology in order to highlight the role of the cell surface in development. Consequently I had to be selective in the choice of suitable examples of developing systems for the discussion. This might have left many lacunae in the citation of contributions from various workers. In general the bibliographic references are intended to be informative to the reader rather than to give credit to individual scientists who published the original reports. Thus not all publications on a given finding are cited. I apologize to the scientists whose important work, in spite of being relevant, is not mentioned in my narration.

I owe a great deal of thanks to several persons for different reasons. My wife, Sunanda, has been an unending source of inspiration in all my efforts. I thank her for bearing with the various problems which arose during the preparation of this book. I thank my sons, Vinay and Ashok, who helped me with enthusiasm in comparing the final typescript with the original draft. Dr. Shashanka Bhide deserves special thanks for his help in comparing the final typescript. I take this opportunity to thank my teacher, Professor Leela Mulherkar, who introduced me to this fascinating subject. In the preparation of the illustrative material, I received help from Mr. R. K. Bhandari and Mr. E. A. Daniels. Mr. K. V. L. N. Rao, the administrative officer of the department, has rendered help in various ways. Mr. K. R. Sharma, the librarian of the department, deserves a special mention for help in locating a number of references. He often pursued the matter even after I had given up. Mr. Baldev Singh Rana has exhibited remarkable patience in typing the manuscript. I am thankful to all these persons.

I am greatly indebted to Dr. G. V. Sherbet and Prof. Ruth Bellairs who read some parts of the manuscript. Their criticism and suggestions have contributed to improving the text considerably. In spite of the help obtained from them, however, I own complete responsibility for any errors and omissions in the book. It was a pleasure to work in association with Mr. B. J. Starkoff of CRC Press and his colleagues during the production of the book. I am thankful to the authors and publishers of various books/journals from whom I have borrowed material. These are acknowledged individually wherever such contributions have been used.

My research efforts in the past two decades have been aided financially by various agencies including the Council of Scientific and Industrial Research, The Indian Council of Medical Research, the Indian National Science Academy, and the Department of Science and Technology, Government of India.

Delhi
September 1985   **K. Vasudeva Rao**

THE AUTHOR

Dr. K. Vasudeva Rao is and has been a teacher and researcher in the University of Delhi where he has given developmental biology courses at the undergraduate and Master's level for more than twenty years. He is especially interested in testing and evaluating material for introducing new exercises in developmental biology for undergraduate and Master's courses.

Dr. Rao started his research as a student of the eminent Indian embryologist Prof. Leela Mulherkar at the University of Poona. After obtaining his Ph.D. degree, he moved to the University of Delhi as a lecturer. Since 1983 he has been a Professor. He has published more than forty papers and three reviews. His current research programs are intended to elucidate the cellular basis of morphogenetic changes. Dr. Rao's research has received support from the Indian National Science Academy, New Delhi, and the central government funding agencies such as the Council of Scientific and Industrial Research, the Indian Council of Medical Research, and the Department of Science and Technology (Government of India).

Dr. Rao is a member of the Indian Science Congress Association, the Indian Society of Cell Biology, and the Indian Society of Developmental Biologists.

With grateful thanks and respect to my teacher

PROFESSOR LEELA MULHERKAR

TABLE OF CONTENTS

Volume I

Chapter 5

CELL FUSION IN DEVELOPMENT

I. INTRODUCTION

As a structural unit of multicellular organisms, the cell can be considered as a nucleated mass of cytoplasm surrounded by a plasma membrane. Within the cytoplasm are elaborated the macromolecules that define the differentiated state of the cell. Many diverse experimental studies have established the fact that the synthesis of substances within the cell is controlled largely by the genetic information in the nucleus. External stimuli such as embryonic inducing substances, hormones and chalones, also regulate the synthetic activities. Some elegant experiments using the unicellular alga, *Acetabularia*, have shown that the development of the phenotype is determined largely by the nucleus[1].

Studies on amphibian embryos using the technique of nuclear transplantation, on the other hand, show that the nuclear activity is also under the influence of the cytoplasm.[2] It is evident from the currently available data that the development of cell phenotype is the result of nucleocytoplasmic interactions as well as the influence of the cellular and noncellular environment. In other words, differentiation is controlled by genetic as well as epigenetic factors. The latter include the environment of the cell. Cell culture is a powerful tool to investigate problems concerning the interaction between a cell and its environment. However, nucleocytoplasmic interactions could be studied only after the somewhat sophisticated techniques of nuclear transplantation and cell fusion were developed. Experimental fusion of two cells offers an opportunity to study the expression of genetic information in a new cytoplasmic milieu. Cell biologists have demonstrated that a variety of cells can be fused forming hybrid cells. Even plant-animal cell fusion has been achieved.[3] However, most cells do not fuse readily. Even tissue cells with intimate junctional complexes such as gap junctions do not fuse. Successful fusion of certain cells has been possible only by employing highly artificial methods such as the use of inactivated Sendai virus, lysolecithin, or polyethylene glycol as fusogens.[4-7] Recently, a novel method of cell fusion has been described by Zimmermann and Vienken.[8] This method dispenses with the use of fusogenic viruses and chemicals. It consists of exposing cells to an alternating, nonuniform electric field of low strength followed by the application of a short electric field pulse of high intensity. This results in cell fusion. For a general account on somatic cell hybridization and genetic analysis of the hybrids, see Evans et al.[9]

A recent biological technique involving cell fusion must be mentioned because of its numerous possible applications. It is the production of monoclonal antibodies. Even when a very highly purified antigen is used for immunizing an animal, the antibodies developed in its blood are heterogeneous. This is due to the fact that many different immune cell clones are formed, and they secrete different kinds of antibodies that recognize different antigenic determinants of the injected antigenic molecule. In 1975, Kohler and Milstein[10] described a method by which the clone of a single immune cell can be amplified so that antibodies with specificity for a single antigenic determinant can be produced. These are called monoclonal antibodies. An important aspect of this technique is the fusion of the immune cell with another of an established cell line so that the hybrid cell (hybridoma) continues to divide and secrete the monoclonal antibodies. Another interesting application of the technique of cell fusion is the microinjection of substances into living cells. The substance to be injected can be held in erythrocyte ghosts, which in turn can be fused with another cell. For a review on this technique and its applications, see Furusawa.[11]

As suggested above, the techniques of experimental cell fusion are important in diverse areas of biological investigation. There are, on the other hand, some developmental processes in which cell fusion occurs as a normal event, and therefore considerable interest centers around them. What can be achieved only by the application of highly artificial means occurs routinely in these developmental processes. Eggs and sperm fuse readily, provided that they belong to the same species and other conditions are optimal. The skeletal muscle is another example of a developmental process involving cell fusion. As a somewhat rare event, cell fusion occurs even in the adult. Multinucleate giant cells derived from macrophages are commonly found in inflammed tissues such as granulomas. Sites of injection of Freund's adjuvant commonly show giant multinucleate cells. Even an inert object such as a coverslip implanted subcutaneously elicits the fusion of macrophages.

The origin of the giant cells from macrophages has been demonstrated experimentally. When ³H-labeled cells are injected, the giant cells show labeled nuclei. Circulating monocytes sometimes form multinucleate cells known as osteoclasts.[12,13] The nuclei within these cells do not divide. Such cells are specially prominent in the giant cell tumor of bone, known as osteoclastoma. Another notable example of multinucleate cells is the syncytiotrophoblast of the mammalian placenta.[14,15] Other instances of natural cell fusion include the formation of slime mold plasmodia and fungi. Plant protoplasts (plant cells denuded of their cellulose walls) can be induced to fuse in the presence of fusogens such as polyethylene glycol. Howver, they sometimes fuse during preparation without the use of any fusogen.

It is intended here to examine the role of the plasma membrane in cell fusion. Only fertilization and myoblast fusion will be dealt with since considerable information is available on these. Besides, these are "developmental" changes and hence of direct interest in the present book.

II. FERTILIZATION

Essentially fertilization consists of the fusion of two gametes, viz., the egg and the spermatozoon. These are highly specialized cells and have undergone considerable differentiation preparatory to the final step, viz., fusion. In mammals, the sperm differentiate in the seminiferous tubules and pass through the epididymis before being ejaculated. After ejaculation, they pass through the female genital tract before meeting the egg. During their complex itinerary, the sperm respond to the changing environment and undergo important modifications. A general strategy of fertilization is that until the sperm reach the egg, they should be able to conserve their energy and be able to attach specifically to the egg or its covering. Equally important is that they should not adhere to any other cell or extracellular material en route. The changes undergone by the sperm during their passage from the seminiferous tubules to the site of fertilization (generally the oviduct) are all geared to fulfill this requirement. Even in animals that shed their gametes into the aquatic environment, the eggs and sperm are specialized to accomplish fusion in the most efficient manner. Capacitation and acrosome reaction are important prerequisites of the essential process of fertilization, viz., cell fusion. The sperm do not undergo these changes until they are close to the egg. The egg and its coatings also play a definite role in the fulfillment of the general strategy of fertilization. In those animals in which fertilization is external, there are special strategies to ensure the success of the event. Specialized spawning behaviors result in the males and females coming together so that the eggs and sperm are released in close vicinity of each other. Though these phenomena have been recognized for a long time, the molecular mechanisms are just beginning to be understood.

During fertilization, there is a coordinated sequence of membrane fusions. The first of these constitutes the acrosome reaction wherein the membrane of the acrosomal vesicle fuses with the plasma membrane. The second membrane fusion is the one involving the plasma membranes of the egg and spermatozoon. In the eggs of many animals, there are a number of membrane-limited vesicles derived from the Golgi. These are closely associated with the egg cortex, and on egg-sperm fusion, they fuse with the plasma membrane, releasing their contents. This constitutes the cortical reaction. Finally, after the spermatozoon has entered the egg cytoplasm, the two pronuclei fuse involving the nuclear membranes. Considerable effort has gone into elucidating the process of membrane fusion, especially that at the cell surface.

A. Surface Specializations of the Sperm

During the differentiation of the spermatids, many important changes must be taking place on their surface as suggested by the overall change in morphology as well as their release from the intimate contact with the Sertoli cells.[16] It is interesting to note that this is a case of haploid cell differentiation. Another point of interest is that this differentiation occurs after the developing germ cells have crossed the blood-testis barrier and have developed a special association with the Sertoli cells. Within the protected environment of the blood-testis barrier, sperm-specific molecules foreign to the male can be synthesized and inserted into the plasma membrane. These surface specializations render the differentiated spermatids autoantigenic. Distinct autoantigens have been located over the acrosomal and postacrosomal regions of the sperm.[17] The distribution of the surface molecules (integral as well as peripheral) seems to change continuously.[18,19] Specific staining techniques have been used to reveal the elaborate specialization of the sperm surface in an attempt at elucidating the functional significance of the chemical entities. A number of antigens are associated with the sperm surface as revealed by the technique of staining with fluorescent antibodies. Some of these antigens seem to originate from the seminal fluid.

Regional differences in the distribution of the surface antigens has been demonstrated in case of human spermatozoa. These studies have evinced the interest of reproductive biologists since the information is expected to be of value in the treatment of infertility caused by autoimmune reactions, and in developing immune reactions against spermatozoa for the purpose of contraception. Some sperm specific antibodies bind exclusively over the post-acrosomal region. Using other methods of visual demonstration, considerable regional differentiation of the sperm surface has been revealed. Colloidal iron particles bind only the periphery of rabbit sperm acrosome. Regional distribution of some membrane enzymes has also been demonstrated.[20] It has been reported that some sperm specific antigens are highly conserved during evolution. Lopo and Vacquier[21] have reported that an antiserum raised against whole glutaraldehyde-fixed sperm of the sea urchin *Strongylocentrotus purpuratus* cross-reacts with the surface of spermatozoa of 28 species of animals representing seven phyla of the animal kingdom (see Table 1). Adequate control experiments were performed by Lopo and Vacquier[21] to rule out the possibility that the antigenicity is obtained as an artifact. Thus the possibility of the male specific H-Y antigens, products of the genital tract and other possible unspecific antigens being detected by the antiserum was ruled out, proving definitively that the antigens detected are associated with the sperm surface. Such a highly conserved feature cannot be without a definite and important function. However, nothing is known about the biological role of these sperm-specific antigens.

The use of lectins as plasma membrane probes has been discussed in an earlier chapter. Occurrence of lectin binding sites on the surface of rat and mouse sperm has been shown by their agglutination in the presence of Con A. Lectin binding studies reveal considerable heterogeneity in the distribution of the receptors on the sperm surface. Lectins with a carbohydrate specificity similar to that of Con A (α-D-glucose and α-D-mannose) show quite different labeling patterns. Whereas Con A binds primarily on the acrosomal region, green pea lectin binds on the nonacrosomal region and lentil lectin binds all over the surface.[18] It must be pointed out in this context that lectin binding specificity is not simply determined by the monosaccharide. Factors such as steric features determining the accessibility of the

Table 1
REACTION OF SSA PREPARED AGAINST
STRONGYLOCENTROTUS PURPURATUS SPERM
(SEA URCHIN) WITH SPERM OF OTHER SPEICES

Phylum Coelenterata	Phylum Echinodermata
Class Anthozoa	Class Echinoidea
Metridium senile fibriatum	*S. purpuratus*
Phylum Annelida	*S. franciscanus*
Class Polychaeta	*S. pallidus*
Fam. Spionidae (1 species)	*S. droebachiensis*
Fam. Polynoidae (1 species)	*Lytechinus pictus*
Phylum Mollusca	*Arbacia punctulata*
Class Amphineura	*Tripneustes gratilla*
Cryptochiton stelleri	*Dendraster excentricus*
Class Gastropoda	Class Ophiuroidea
Acmaea sp.	*Ophioplocus esmarkii*
Class Pelecypoda	Class Asteroidea
Macoma nasuta	*Patiria miniata*
Phylum Echiuroidea	Phylum Chordata
Urechis caupo	Class Urochordata
Phylum Arthropoda	*Styela clava*
Class Crustacea	*S. plicata*
Cancer antennarius	*Ciona intestinalis*
Pinnixa tubicola	Class Osteichthyes
Class Merostomata	*Salmo* sp.
Limulus polyphemus	Class Amphibia
	Rana pipiens
	Class Aves
	Meleagris gallopava
	Class Mammalia
	Mesocricetus auratus
	Rattus norvegicus

Note: Reactivity of the sperm specific antiserum (SSA) was assayed by indirect immunofluorescence or by the immunoperoxidase procedure. The sperm were fixed in either 3% glutaraldehyde or formaldehyde. Preimmune serum did not react.

binding sites and the complexity of the cell surface oligosaccharides bearing the specifically binding monosaccharide also have a role.

In view of this, it may be surmised that the surface of the sperm has an elaborate organization of the plasma membrane presumably related to its function. Nicolson and Yanagimachi[22] studied the pattern of agglutination of sperm caused by different lectins. Head-to-head agglutination of rabbit epididymal spermatozoa was caused by Con A, WGA, and RCA, whereas hamster spermatozoa exhibited tail-to-tail agglutination in the presence of these lectins. From this, it may be concluded that considerable degree of species specificity exists with regard to the location of the lectin binding sites. Nicolson and Yanagimachi[22] observed that ferritin coupled RCA (which has β-D-galactose specificity) binds uniformly over the surface of rabbit spermatozoa when exposed to the lectin at 0°C. When such cells were washed and incubated at 37°C, clustering of ligands was observed, indicating that the ligands can move laterally within the plane of the plasma membrane. In other words, the ligands are integral membrane components. Various elements of the cytoskeleton have been implicated in the maintenance of the surface ligands.[23] Lectin binding and agglutination of

Table 2
TISSUE SPECIFICITY OF
MONOCLONAL ANTIBODIES

Absorption[a]	Antibody[b] titer^{-1}		
	J1	**C6**	**A5**
Nil	512	256	512
Brain	256	256	512
Liver	512	256	512
Kidney	512	128	256
Spleen	256	128	512
Testis	0	0	0
F 9 EC	512	2	0

[a] One volume of hybridoma supernatant was mixed with 2 vol of minced tissue on ice for 3 hr.
[b] Data represents the last doubling dilution of hybridoma supernatant giving a color reaction above background in an enzyme immunoassay using adult testicular cells as targets.

From Fenderson, B. A., O'Brien, D. A., Millette, C. F., and Eddy, E. M., *Dev. Biol.*, 103, 117, 1984. With permission.

sperm shows that glycoproteins and glycolipids are important components of the plasma membrane. The existence of sialic acid as one of the sugars in these molecules is suggested by the observation that neuraminidase-treated sperm show an approximately 30% decrease in their electrophoretic mobility. Sialic acid is present in the epididymal lumen, and it has been suggested that sialic acid-associated components on the spermatozoa vary according to their maturity.[24,25]

Considerable work has been done on the surface electrical properties of the sperm. Most of this was occasioned by the desire to separate X and Y chromosome-bearing sperm to control the sex ratio in the progeny of animals after artificial insemination. No success has been reported in this regard. However these studies have revealed that the sperm of different species vary widely in their net surface charge.

Recent studies have employed the technique of raising monoclonal antibodies against sperm-specific antigens. Using these as detecting agents, different germ cell surface antigenic determinants have been reported. Binding of the monoclonal antibodies is revealed by indirect immunofluorescence. Fenderson et al.[26] obtained three monoclonal antibodies against different sperm-specific antigens. These were designated J1, C6, and A5. The reactivity of the surface to these antibodies changes during sperm differentiation (Table 2). Reactivity with epididymal cells shows that both J1 and C6 specific determinants are present over the anterior portion of the acrosome where peanut agglutinin also binds. However, the antibodies do not block the binding of peanut agglutinin. Fenderson et al.[26] have provided evidence that J1, C6, and peanut agglutinin identify different sperm surface antigens. The carbohydrate nature of the antigenic determinants has been shown by the observation that binding of the monoclonal antibodies is inhibited by specific mono- and disaccharides.

B. Capacitation and Acrosome Reaction

Mere morphological differentiation of the mammalian secondary spermatocytes into spermatids does not mark the end of changes necessary for fertilization. The sperm undergo a series of physiological changes. Sperm from the caput or cauda epididymis are not able to fertilize efficiently. Definite sperm maturation changes in the plasma membrane have been demonstrated to occur in the epididymis.[27,28] Even sperm from vas deferens or semen need to pass through some change before they can undergo the acrosome reaction. It appears that this change, known as capacitation, occurs in the female genital tract. Something resembling capacitation is known even in case of Anura, where external fertilization occurs. Exposure of the sperm to the egg jelly substances brings about the change. Dejellied eggs are not fertilized efficiently. Capacitation was first described by Austin[29] and Chang.[30] The essential requirement of a few hours' residence of the sperm in the female reproductive tract, before acquiring fertilizing ability, has been recognized.[31,32] Removal of some surface components seems to be an essential part of the process. The substances removed from the sperm surface seem to be proteins/glycoproteins. There is another series of changes at the sperm surface associated with capacitation: the ratio of phospholipid to cholesterol changes.

As mentioned above, capacitation is a physiological change undergone by the sperm in response to the fluids in the female genital tract. The change includes, at least in part, the loss or dissociation of some seminal fluid components associated with the sperm surface. However, capacitation is not merely a loss of seminal fluid components. Even sperm from caput or cauda epididymis need to undergo capacitation. The surface changes accompanying capacitation are just beginning to be understood. Early studies by Gordon et al.[33,34] on the pattern of Con A binding on spermatozoa obtained from different parts of the male and female genital tracts have indicated the complex nature of the capacitation changes. Caput sperm showed very little binding, whereas the cauda sperm had a uniform dense binding. This indicates that lectin binding sites are developed during the transit from caput to cauda epididymis. Sperm from the semen before deposition into the female genital tract showed no binding, suggesting that some constituents of the seminal fluid mask the lectin binding sites. Whether this masking has any functional significance is not easy to prove. Capacitated sperm (obtained after deposition into the female genital tract) generally showed reduced binding over the acrosome, but retained the dense binding over the postacrosomal surface. These studies suggest that the appearance and distribution of Con A binding sites on the surface could be related to the functional changes undergone by the sperm.

Ward and Storey[35] have recently described a method to monitor the time course of capacitation changes. Mouse sperm taken from the epididymis are unable to bind the zona pellucida of the egg; they can bind only after the capacitation changes have occurred. If the epididymal sperm are cultured in vitro in the presence of 2% bovine serum albumin (BSA), they undergo changes analogous to those of capacitation. During the course of these changes induced in vitro, a characteristic alteration occurs in the pattern of fluorescence in the presence of chlortetracycline. The sperm from cauda epididymis show a uniform fluorescence over the head with a bright band over the equatorial region. The mid piece also shows bright fluorescence. During culture in vitro in the presence of BSA, the sperm head shows a change: the fluorescence becomes more uniform and eventually it remains only over the anterior part of the head region. This takes about 90 min. Acrosome-reacted sperm show no fluorescence over the entire head (Figure 1). Sperm showing the pattern in Figure 1C could undergo acrosome reaction spontaneously or by treatment with acid solubilized zona material, whereas those with the pattern in Figure 1A could not be induced to undergo acrosome reaction. Thus the change in the pattern of fluorescence during the 90 min in vitro seems to correspond to the capacitation changes. The time taken for this is similar to that taken by mouse sperm to undergo capacitation as shown in other studies. Though the nature of these changes is not clear, the fluorescence method forms a convenient assay.

Do the capacitation changes enable the sperm to recognize the different anatomical parts through which they pass and to respond accordingly? If so, what is the nature of these

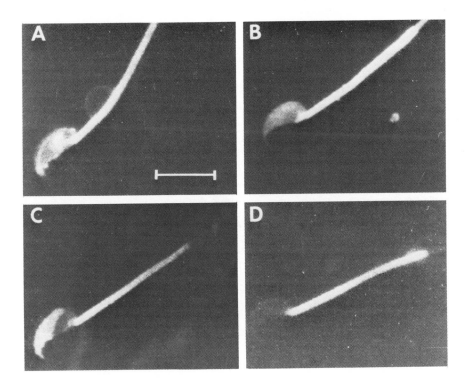

FIGURE 1.   Changing pattern of fluorescence on the mouse sperm surface due to chlortetracycline during capacitation. (A) Pattern showing fluorescence over the head with a bright band of fluorescence across the equatorial segment; (B) uniform fluorescence over the head; (C) fluorescence on the anterior portion of the head and a dark band over postacrosomal region; (D) barely detectable fluorescence over the head of a sperm that has undergone acrosome reaction. Measuring bar, 10 μm. (From Ward, C. R. and Storey, B. T., *Dev. Biol.*, 104, 287, 1984. With permission.)

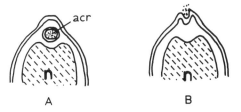

FIGURE 2.   Acrosome reaction. The acrosome vesicle (acr) membrane becomes continuous with the sperm head plasma membrane. The contents of the vesicle are released. (A) Before, and (B) after the reaction. n = Nucleus.

responses? Does the altered surface help in specific adhesion of the spermatozoon to the egg barriers in preference to the lining of the genital tract? These questions, though not yet fully answered, are now in the realm of tangible experimental approach.

Most animal sperm have an acrosome at the tip of the head. Even the amoeboid sperm of nematodes have the acrosome, albeit in the form of several acrosome-like bodies. The acrosomal reaction consists of fusion of the acrosomal vesicle with the plasma membrane (Figure 2) and release of the acrosomal enzymes, which help the sperm in passing through the egg barriers. Hyaluronidase and a proteolytic enzyme named acrosin are the best known enzymes associated with the acrosome. It is fairly certain that these lytic enzymes are released from the acrosome vesicle during the acrosome reaction. Hyaluronidase lyses some of the components of the extracellular matrix holding the follicular cells (corona radiata) and enables the passage of individual spermatozoa between the cells. Acrosin causes lysis of the zona pellucida enabling the sperm to approach the egg surface. The proteolytic nature of acrosin has been elegantly demonstrated by its localized action on gelatin. The lytic action is seen only around the sperm head. In the unreacted mammalian sperm, the acrosin exists as a zymogen, proacrosin.[36]

Essentially, the acrosome reaction is a process of exocytosis wherein the membrane of the acrosomal vesicle fuses with the plasma membrane, and thus the contents are released (Figure 2). Attempts have been made to discover the mechanism(s) responsible for triggering the reaction. Some 30 years ago, Dan[37] demonstrated that sea water in which eggs have been kept for some time can trigger the acrosome reaction in star fishes. In some animals, actual contact between the sperm and the vitelline coat is essential for triggering. Studies on the source of the triggering material generally indicate that it is some component of the egg membrane, and it can be extracted by various methods. Some attempts have been made to isolate the triggering agent, but it appears that it is not easy to identify it, especially since Ca^{2+}, another essential factor in the process, can by itself trigger the reaction under certain conditions. Besides, a variety of ionophores including A 23187, X 537a, and nigericin can trigger the reaction.[38-40] It is now established that Ca^{2+} plays an essential role in the acrosome reaction in echinoderms as well as mammals.[36] Another aspect of the acrosome reaction is the polymerization of actin in the periacrosomal region, forming an extruded acrosomal process.[41] It appears that the initial polymerization of the acrosomal process pushes it through the acrosomal vesicle, the extending process thus getting coated with the exteriorized acrosomal contents.[42]

C. Surface Specializations of the Egg

The remarkable species specificity of fertilization suggests the existence of a positive recognition system that allows the binding of sperm of the egg's own species to the exclusion of all other. Eggs are enveloped with more or less elaborate coverings, which could presumably serve to recognize the sperm specifically. There are additional egg coverings of diverse nature serving a nutritive, protective, or other role. Embryologists have recognized the egg coverings under three categories: (1) primary egg membranes, which are secretion products of the egg (the vitelline membranes in insects, annelids, mollusks, amphibians, reptiles, and birds; zona pellucida of mammals; the so-called chorion of tunicate and fish eggs); (2) secondary egg membranes, which are cellular or acellular products of the follicle (corona radiata of mammals, chorion of insects, etc.), and (3) tertiary egg membranes, which are secretion products of the oviduct and other accessory parts of the genital organs (the albumen and calcareous or leathery shells of reptiles and birds; jelly of amphibian eggs; the egg capsules together with the enclosed glycoproteins as in the case of many mollusks). The tertiary egg membranes are generally laid down after fertilization. In general, the primary egg membranes are of a glycoprotein nature though they may contain other material.

In the sea urchins, the egg has a thin vitelline membrane that is intimately associated with the plasma membrane, conforming to the outer surface of the egg that has microvillous projections. The vitelline membrane can be removed by the application of disulfide-reducing agents such as dithiothreitol. Chemically it is a glycoprotein. Sperm mixed with isolated vitelline layer attach only to the outer surface of the latter, suggesting structural asymmetry and the existence of sperm binding sites. Trypsinization decreases fertilizability of the eggs, perhaps by removing the sperm binding sites. In most cases, sperm binding is species specific. Membrane fractions from lysates of unfertilized eggs inhibit fertilization by competitive inhibition.

These and other observations indicate a positive role for the vitelline layer in the egg-

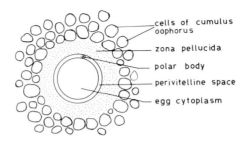

FIGURE 3.   Structures found with the ovulated mammalian egg.

sperm recognition system.[43] The egg surface is produced into fine microvilli, often in patterns characteristic of various species.[44] Regional differences in the density of such microvilli over the egg surface have also been observed in case of *Xenopus*.[45] Vacquir[46] has described a method for isolating the surface of an unfertilized sea urchin egg. The method consists of affixing the eggs to cationic surfaces and shearing them off in a calcium-free environment. Both faces of the surface layer can then be studied by scanning electron microscopy permitting visualization of the surface before or during sperm entry.

Mammalian eggs also have microvilli.[47] However, the role of these structures in the binding of sperm is not known. In mammals, the egg investments are complex (Figure 3), and they are largely responsible for the prevention of interspecific fertilization. Lectin binding characteristics of eggs have been studied with the intention of locating functionally specialized areas bearing glycoproteins.[48] In general, however, it appears that the mammalian egg surface glycoproteins are not involved in fertilization as sperm binding/recognition sites.[48,49] On the other hand, the zona is directly involved in the recognition of the sperm. This does not exclude the possibility of an active role for the egg surface. Saling et al.[50] have shown that a monoclonal antibody against mouse sperm blocks specifically at the level of sperm-egg surface interaction. As in many other animal groups, prevention of interspecific cross-fertilization in mammals occurs at different levels. Besides the behavioral preliminaries to mating and ejaculation, the absence of which could preclude cross-fertilization, there are physiological peculiarities offered by the female genital tract affecting the motility and even viability of the sperm that may be deposited experimentally by artificial insemination.[51] Even the complex structures around the egg could prevent further advances of the sperm, which may have reached the upper part of the oviduct and come in the vicinity of the ovulated egg.

In particular, the zona pellucida seems to be the most effective barrier, as shown by the successful fusion of sperm from widely unrelated species when in contact with zona-free eggs. For example, human spermatozoa can undergo acrosome reaction, fusion, and pronucleus formation by interacting with zona-less hamster eggs. A zona-free egg is, however, not necessarily ''open house'' for all kinds of sperm; human sperm cannot bind rat or mouse oocyte after removal of the zona. Capacitated human sperm do not bind the zona of the baboon oocyte, though they attach avidly to the zona of the gibbon and penetrate it (Figure 4). Obviously many things are murky in this field.[51] However, the target of future work is clearly defined, and it is the surface carbohydrates of the gametes.[52]

D. Sperm-Egg Fusion

A firm binding of the sperm on the egg surface is essential for the central event of fertilization, viz., fusion of the gametes. In sea urchins, it has been shown that a specific binding protein, *bindin* derived from the acrosome reacted sperm, ensures such a firm adhesion. This protein has special affinity for the egg surface as inferred from the fact that it causes agglutination of eggs species specifically.[43,53] Methods for the purification of

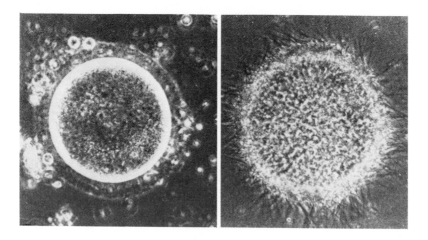

FIGURE 4.   Specificity in egg-sperm interactions. A baboon oocyte (left) and a gibbon oocyte (right) were exposed to similar concentration of human sperm. Capacitated or not, the human sperm show no affinity for the zona pellucida of the baboon egg. They, however, adhere avidly to the zona of gibbon. (Reprinted by permission from *Nature*, 291, 286. Copyright © 1981, Macmillan Journals Limited.)

bindin[54] and an assay for its release[55] have been described. Attempts have been made to identify the sperm-receptive substances on the egg surface. A factor obtained from the egg membrane of *Arbacia* can inhibit fertilization. This factor is trypsin sensitive.[56] The factor has affinity for Con A, and the latter itself can inhibit fertilization. These early studies indicated that the egg surface carbohydrates could play the key role in species-specific recognition of the sperm. Recent studies have amply vindicated this suggestion. Besides, they show that the general nature of this mechanism has been conserved during evolution. Ahuja[57] demonstrated that binding of capacitated hamster spermatozoa to the zona pellucida can be inhibited by several monosaccharides and oligosaccharides related to fucose and acetylated amino sugars. Several glycoproteins with terminal galactose-*N*-acetyl glucosamine were also potent inhibitors. No inhibition could be obtained in the presence of other unrelated sugars. Treatment of the capacitated spermatozoa with α-L-fucosidase, α-D-galactosidase and β-*N*-acetyl hexosaminidase could effectively inhibit fertilization. However, other glycosidases and arylsulfatases did not have inhibitory action. These observations clearly implicate sperm surface ligands containing fucose, galactose-*N*-acetylglucosamine, and *N*-acetylgalactosamine residues in sperm-zona binding. Huang and Yanagimachi[58] have demonstrated that fucoidin inhibits the attachment of guinea pig spermatozoa to the zona pellucida.

   Work on sea urchins has now clearly established the participation of surface sugars in species-specific recognition of the sperm and egg surfaces. The mechanism consists of bindin and complementary egg surface molecules. The egg surface ligands of two sea urchin species, *Strongylocentrotus purpuratus* and *Arbacia punctulata*, have been investigated.[59] The ligands bind spermatozoa species specifically only after acrosome reaction, clearly indicating that the sperm and egg surfaces have complementary molecules. Eggs exposed to antibodies against the egg surface ligands are not fertilizable. Rossignol et al.[59] have isolated the egg surface ligand and shown that it is a glycoconjugate of very high molecular weight ($>10^6$) and shows some properties of proteoglycans. The isolated egg surface ligands bind the sperm species specifically. The protein component of the egg surface receptor of *S. purpuratus*, though not a single polypeptide, is not a very complex mixture either. Only four N-terminal amino acids could be detected. The glycan component is very large ($\approx 10^6$ mol wt) containing iduronic acid, galactosamine, and sulfate. The high degree of sulfation confers a high degree of negative charge on the proteoglycan complex. It has been shown that it is the carbohydrate

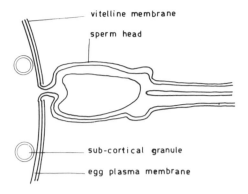

FIGURE 5.   Fusion of the spermatozoon with the egg surface establishing
cytoplasmic continuity between the two.

parts and not polypeptides of the surface receptors that are responsible for species-specific recognition.

In some animals, the point of entry of the sperm bears a definite relation to the animal-vegetal axis of the egg, and this in turn seems to influence the axial organization of the embryo developing from it.[60] The distribution of the sperm receptors may in such cases determine the point of entry of the sperm. However, more information is needed to confirm this. Vertebrate eggs have numerous microvilli over their surface, and it has been suggested that the spermatozoon first makes a contact with a microvillus.[61]

The process of membrane fusion during fertilization has been studied extensively in the invertebrates and, in particular, in echinoderms. In these forms, the remaining inner membrane of the reacted acrosome fuses with the egg plasma membrane (Figure 5). Further progress of the spermatozoon into the egg cytoplasm depends on contractile proteins.[62] Initial fusion of the mammalian egg does not seem to involve the inner acrosomal membrane. On the contrary, it is the membrane over the equatorial region of the sperm head that seems to be involved. Considerably detailed information on the early membrane changes during fertilization has been revealed by electron microscopic studies, especially using eggs fertilized in vitro (Figure 6).[63] When the sperm approaches the egg surface, the first contact established is between the equatorial surface of the sperm head and the egg plasma membrane where fusion occurs over a small region. Following this, the membrane over the postacrosomal surface regresses or peels back, and the egg cytoplasm flows over the postacrosomal surface of the nucleus. Small vesicles (Figure 6) probably derived from the fusion of sperm and egg surfaces are seen in the cytoplasm at the point of initial fusion. In the meanwhile, the acrosomal region (covered by the inner acrosomal membrane, which is exposed during the acrosome reaction), is enveloped by the egg plasma membrane. As a result, a vesicle consisting of the acrosomal inner membrane and the endocytosed egg plasma membrane is formed around the rostral end of the spermatozoon. An important point that needs further elucidation is the mechanism that promotes membrane fusion. Obviously, the lipid layers have to be destabilized so that the cytoplasm of the two cells can become contiguous. Conway and Metz[64] have observed that a phospholipase appears shortly after the acrosome reaction in the sea urchin sperm. It has been suggested that the enzyme destabilizes the lipid membranes and thus initiates fusion.

E. Polyspermy and Its Prevention

We have emphasized earlier that the sperm and eggs are highly specialized cells and possess mechanisms that enable them to fuse with each other. However, there are special controls to prevent them from going too far. In the preceding sections, we have discussed

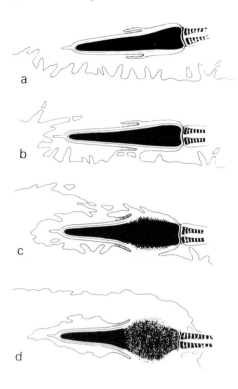

FIGURE 6.   Diagrams to show the sequence of events during the fusion of a mammalian spermatozoon with the egg. The sperm head is shown in association with the villous egg surface (represented by the irregular line). (a) Prior to establishing contact with the egg surface; (b) an initial stage of fusion with a small, defined region of the equatorial segment of the sperm head fused with the egg surface; (c) and (d) further progress of the sperm into the egg cytoplasm by a process of peeling back of the membrane of the postacrosomal surface. The egg surface also covers over the membrane limited acrosomal region. Progressive decondensation of chromatin is also shown in the sperm nucleus. (From Bedford, J. M. and Cooper, G. W., *Cell Surface Rev.*, 5, 65, 1978. With permission.

the evidence that special cell surface molecular mechanisms ensure that the sperm or eggs do not fuse with any other type of cells but only with each other of the same species. There is another restriction on the fusion of these cells. Normally once an egg is fertilized, it does not fuse with other sperm. It quickly develops some changes in the plasma membrane preventing the entry of additional sperm. The membrane changes following fertilization seem to extend over the egg surface in a wave-like manner, starting from the point of entry of the successful spermatozoon. Larger eggs (e.g., those of birds) normally allow the entry of several or many sperm, presumably because the membrane changes do not reach the entire surface area of the egg rapidly enough. However, even in these cases, only one spermatozoon forms the male pronucleus. This view finds further support from the observation that even in small eggs (in which the postfertilization surface changes spread rapidly over the entire egg), polyspermy may occur under experimental conditions of very high sperm densities that enhance the chances of additional sperm fusing with the egg surface before the changes could spread.

The block to polyspermy is some change that either eliminates the sperm receptor sites, or makes them inaccessible. Work on sea urchin eggs indicates that block to polyspermy is developed in two distinct phases and by distinct mechanisms. First, a fast, partial, and temporary block to polyspermy results from a sudden change in the egg membrane potential from − 60 mV to about + 10 mV, which presumably alters the sperm binding sites.[65] There is strong evidence that the change in membrane potential is related to the prevention of polyspermy. Eggs that fail to show depolarization to more positive values become poly-

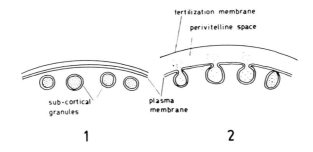

FIGURE 7. Elevation of the fertilization membrane from the egg surface. 1, Diagram of the egg surface before fertilization; 2, expansion of the perivitelline space and elevation of the fertilization membrane after egg-sperm fusion.

spermic. Unfertilized eggs whose membrane potential was experimentally raised, could not be fertilized. Finally, when current was applied during fertilization to maintain the negative value, the eggs became polyspermic. Following the entry of the first sperm, the receptivity of the egg to other sperm decreases, though at a very high concentration of sperm, polyspermy can still occur.[66] Thus the egg remains "unprotected" against high sperm densities until a subsequent block to polyspermy occurs. Nevertheless, the rate of successful entry of a subsequent sperm is substantially reduced after the first spermatozoon has entered. For a full discussion and references to earlier literature, see Nuccitelli and Grey.[67]

The early block to polyspermy is supplemented by other mechanisms, chiefly the cortical reaction. The egg surface undergoes important membrane changes on fertilization. This is inferred from the slow recovery from photobleaching of fluorescence. In most types of eggs, fertilization is followed by a reaction involving the structure of the plasma membrane. It consists of exocytosis of the subcortical granules and hence is called cortical reaction. Whether the cortical reaction precedes or follows sperm-egg fusion depends on the animal type.[36] The release of the subcortical granules seems to have two consequences. First, the plasma membrane becomes a mosaic of old and newly-surfaced membrane. Eddy and Shapiro[68] have estimated that there is a doubling of the surface area of the sea urchin egg as shown by scanning electron microscopy and doubled Con A binding after fertilization. Second, there is the formation of a fertilization membrane. By exocytosis, the subcortical granules are eliminated, and their material fills the space between the egg plasma membrane and the vitelline membrane. Externalization of the cortical granule material and its hydration elevates the vitelline membrane as a distinct fertilization membrane (Figure 7). Although the cortical reaction has been studied extensively only in the sea urchin egg, analogous changes occur in the other eggs also.[69] As a consequence of the cortical reaction, adhesion of sperm on the egg surface is prevented. It must be pointed out that the cortical reaction constituting the "late block to polyspermy" establishes a permanent physical barrier, which is also more effective than the early block.

It is not known if an elaborate mechanism to prevent polyspermy as described above exists in other animals. In the mouse, it has been reported that there is no electrical block to polyspermy.[70] The cortical reaction seems to be triggered by an elevated level of intracellular Ca^{2+} in the egg and does not depend on the extracellular Ca^{2+} level. The essential role of Ca^{2+} has been demonstrated in the eggs of echinoderms, amphibians, and mammals.[71] An impressive piece of evidence for the correlation between the triggering of cortical reaction and elevated level of Ca^{2+} has been provided by Ridgeway et al.[72] who used the eggs of medaka, a teleost fish. Aequorin, a Ca^{2+}-dependent luminescent photoprotein, could be injected into the egg before fertilization. Luminescence of this depends on free Ca^{2+} concentration. Sperm entry is effected through a small orifice, the micropyle, in the thick egg barrier. Before fertilization, the aequorin-injected egg shows a very low resting glow. As soon as the spermatozoon enters, luminescence appears at the micropyle and then extends over the surface of the egg. The elevated luminescence is about 10,000 times the resting level. This indicates that the release of free Ca^{2+} spreads over around the egg starting at the point of entry of the spermatozoon. In sea urchin eggs also, the same technique has been used for a demonstration of the elevated Ca^{2+} level. The source of this Ca^{2+} is intracellular. A calcium binding protein found in the microsomal fraction has been shown to be the source of the rising level following fertilization. It has been indicated that the release of Ca^{2+} is itself Ca^{2+} dependent, and the initial stimulus is provided by the Ca^{2+} released from the sperm during the acrosome reaction.[36,73]

During fertilization, the sperm nucleus and cytoplasmic organelles enter into the egg cytoplasm. What happens to the plasma membrane? There is evidence that the plasma membrane remains on the surface of the fertilized egg, at least initially, thus making the fertilized egg surface a mosaic of the two gamete membranes.[74] This was shown by the observation of FITC-labeled sperm surface component on the fertilized egg in both mouse and sea urchin. The fluorescent patch was internalized,[75] and this internalization was mediated by the cytoskeleton of the egg. A rat sperm surface antigen, which is absent on the surface of the unfertilized egg, can be detected on the egg surface after fertilization. Binding of a monoclonal antibody against the antigen can be demonstrated by indirect immunofluorescence. Further, in the presence of rabbit complement, the monoclonal antibody causes immunolysis of the fertilized egg. The pattern of the lateral spread of the sperm surface antigen indicates that it spreads within the membrane. For details and references, see Gaunt,[76] and Gundersen and Shapiro.[77] Though the incorporation of some sperm surface antigens into the egg surface thus seems to be established, their functional role is still in the realm of speculation.

F. Strategies of Fertilization

In all animals and plants which reproduce sexually, we find an amazing strategy of fertilization, modified accordingly as to whether the process takes place internally or externally. Fusion of gametes from distinct individuals ensures mixing of hereditary material leading to variation. Even in hermaphrodites, self-fertilization (i.e., fusion of gametes from the same individual) is avoided, and cross-fertilization is ensured by various contrivances. Generally, hermaphroditic animals produce gametes of the two sexes at different times so that self-fertilization is avoided. Another essential feature of sexual reproduction is that it should ensure stability of the species. In other words, the gametes of different species should not fuse. In the higher animals, interspecific fertilization is rendered improbable by the mechanical incompatibility of genitalia or owing to differences in the preliminary behavioral patterns that lead to mating. In addition, even if insemination takes place, there are other barriers between the egg and sperm, the most important of them being the adhesive specificities developed on their surfaces.

In plants, the strategies to avoid fertilization of gametes of different species are combined with the contrivances to render self-pollination infructuous. When the pollen grain of the acceptable type (i.e., of the same species but of a different individual plant) falls on the receptive surface of the stigma, it is hydrated. Passage of water through the surface of stigma cells and into the pollen grain is critical for pollination in many flowering plants. A control of this process is known to ensure germination of compatible pollen and failure of incompatible ones. Some stigma surface proteins interact with the proteins of the pollen wall. In case of compatible pollination, the pollen grain germinates, sending out a tube, the pollen tube, which enters the stigma and grows through the style. Eventually, the pollen tube reaches the ovary and finally the embryo sac. Once the pollen tube has entered the embryo sac, fertilization is ensured (Figure 8). However, barriers in the form of different pistillar

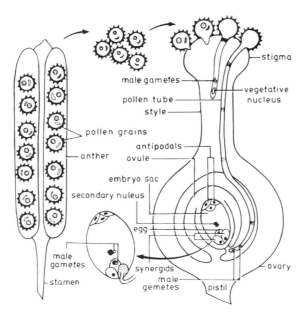

FIGURE 8. Diagrammatic representation of sexual reproduction in flowering plants. Pollen grains, which carry the male gametes or their progenitor cells, are produced in the anthers and become deposited on the stigma through various agencies. The pollen grain germinates on the stigma and produces a pollen tube, which grows through the tissues of the stigma and style and reaches the embryo sac, the container of the female gamete. The pollen tube enters the embryo sac through one of the synergids and discharges the two male gametes. One of the male gametes eventually fuses with the egg nucleus to produce the zygote and the other with the secondary nucleus to produce the primary endosperm nucleus. (Courtesy of Dr. K. R. Shivanna, University of Delhi.)

tissues can inhibit the germination and growth of pollen tubes in cases of incompatibility. The first interaction leading to rejection or acceptance of the pollen occurs immediately after the pollen grain lands on the stigma. In case of incompatibility, pollen hydration does not occur or the pollen tube fails to enter the stigma tissues. The stigma surface has certain proteins/glycoproteins that are involved in the interaction. Knox et al.[78] demonstrated the presence of Con A binding sites on the stigma surface. Con A binding itself prevents the entry of the pollen tube, but does not affect pollen germination.

However, other stigma surface proteins are required for normal germination of the pollen. In case of interspecific pollination, the stigma surface proteins may be such that no pollen germination occurs or the pollen tube fails to penetrate the stigma tissues. The pollen wall proteins are also involved in this interaction.[79,80] Incompatibility in some cases can be overcome by introducing killed compatible pollen along with the incompatible one,[81] thereby showing that the pollen surface proteins also play a positive role in recognition and eliciting the stigma reaction. The essential role of the stigma-pollen interaction is also demonstrated by experiments in which pollen grains were placed on the cut surface of the style. Both compatible and incompatible pollen germinated, and the tubes entered the ovary. On the other hand, incompatible pollen placed on the stigma germinated, but could not pass through the tissues of the style. Raff and Knox,[82] who made these observations, suggest that the incompatible pollen-stigma interaction brings about some change in the tissue of the style, which prevents the growth of the tube. The nature of this change is uncertain, but it may be concerning the tract geometry.

In some plants, the tissue of the style inhibits the growth of the tube of an incompatible pollen. The different levels at which incompatibility is known to be expressed in different flowering plants and its probable mechanisms are represented in Figure 9. Fertilization involving the gametes of unrelated plants (incompatible pollination) by culturing ovules and

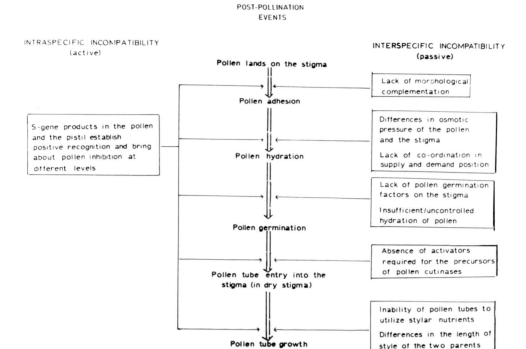

POST-POLLINATION
EVENTS

INTRASPECIFIC INCOMPATIBILITY
(active)

INTERSPECIFIC INCOMPATIBILITY
(passive)

FIGURE 9.   Diagram representing the postpollination events occurring in a typical flowering plant. On the right and left are shown the factors that result in rejection/incompatibility leading to the failure of fertilization. (Courtesy of Dr. K. R. Shivanna, University of Delhi.)

pollen grain in vitro has been achieved.[83] Most of this research was occasioned by the desire to bring about fertilization where normal pollination fails. Incidentally however, it has led to the conclusion that the discrimination between compatible and incompatible crosses is exercised especially by the interactions between the pollen and pistillate tissue. This is reminiscent of the observation that prevention of interspecific fertilization in mammals is exercised by structures residing around the egg (e.g., the zona) rather than by the egg surface.

It does not seem relevant to dwell any further on the subject of pollination in plants. Those interested in the system may refer to the relevant literature.[81,84-88]

III. MYOGENESIS

The skeletal muscle fibers of the vertebrates are multinucleate structures. During development, they arise from the myogenic cells known as myoblasts. Studies on the differentiation of the skeletal muscle have been occasioned largely by the needs of medical scientists who have to understand the causes leading to various states of disease such as muscular dystrophy. The adult muscle is capable of responding to injury, denervation, disuse, and other similar disturbances. The interest of developmental biologists in myogenesis has been chiefly due to two unique features: (1) the skeletal muscle is a highly specialized and terminally differentiated tissue with several muscle-specific products, and (2) it involves cell fusion.

Studies on the development of myotubes from myoblasts has been greatly aided by the facility of direct observation of the process in vitro.[89] When myogenic cells are disaggregated and allowed to settle down to the surface of glass or plastic, they adhere to the substratum. They divide by mitosis and eventually align themselves with their long axes parallel. This

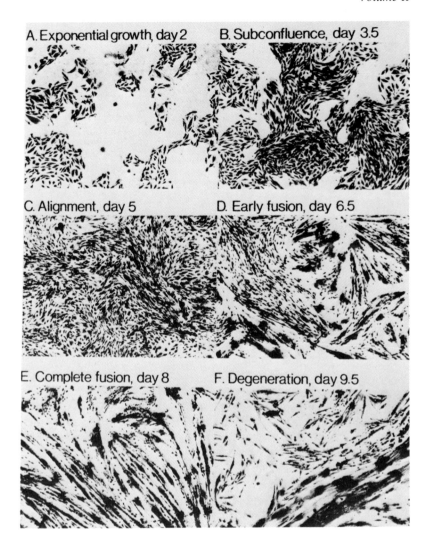

FIGURE 10. Myogenesis in vitro. Formation of multinucleate myotubes from L6 rat myoblasts. Cultures fixed with methanol and stained with Giemsa stain. (From Pearson, M. L.,
in *The Molecular Genetics of Development*, Leighton, T. and Loomis, W. F., Eds., Academic
Press, New York, 1980, chap. 9. With permission.)

alignment is a characteristic feature of the myoblasts that are ready to fuse. The next event
in their development is their fusion to form multinucleate myotubes. The fusion process
continues further, and thus myoblasts continue to fuse with existing myotubes (Figure 10).
The primary cell cultures set up from disaggregated embryonic skeletal muscle include
fibroblasts as well as the myogenic cells. The fibroblasts, however, never align as the fusion
competent myoblasts do; they also do not fuse. Fibroblasts adhere to the glass or plastic
substratum sooner than the myoblasts. This feature permits early harvesting of the nonad-
hering cells and subculturing them as the myoblast-enriched population.

Differentiation of the muscle fibers starts after fusion. The earliest event in the process
of differentiation is a sudden change in the pattern of protein synthesis. The proteins predominantly found in the muscle tissue include actin, myosin, tropomyosin, actinin, creatine
phosphokinase, the acetyl choline receptor, and others. The quantity of these proteins increases enormously after fusion.

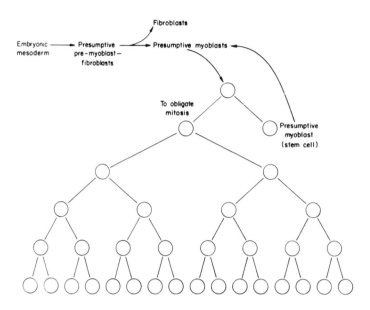

FIGURE 11.   The embryonic origin of fusion competent myoblasts. (Adapted from Quinn, L. S., Nameraff, M., and Holtzer, H., *Exp. Cell Res.*, 154, 65, 1984.)

A. The Origin of Myogenic Cells

Myogenic cells arise from the early embryonic somites, which consist of the dermatome, myotome, and sclerotome. Dividing cells that arise from the myotome are thought to reach the appropriate anatomical sites where myogenesis occurs. Even in more distally located structures such as limbs, the myogenic cells are of myotome origin.[90-93] There is evidence to show the existence of distinct clones of myogenic and chondrogenic cells in the limb-bud mesenchyme. Single cells from early limb bud in the same culture dish yield either chondrogenic-fibrogenic or myogenic-fibrogenic clones.[94] Sasse et al.[95] have shown that myogenic and chondrogenic cells in the early limb-bud mesenchyme can be distinguished by a monoclonal antibody. According to the evidence obtained by Holtzer and co-workers,[96,97] the "presumptive myoblast-fibroblasts" become further committed to form presumptive myoblasts and fibroblasts. Finally, the presumptive myoblasts proliferate, expanding the clone. Myogenic differentiation proceeds when a presumptive myoblast is chartered to undergo an asymmetric division, giving rise to two dissimilar daughter cells, one of which is a presumptive myoblast (stem cell) and the other, a committed cell, destined to undergo four symmetric divisions (see Figure 11 and Quinn et al.[98,99] for further details and references). Having undergone the four obligatory symmetric divisions (giving rise to identical daughters), the cells are fusion-competent postmitotic myoblasts. These can be distinguished from the continuously dividing cells by the detection of the muscle specific isozyme of creatine kinase using an antibody. The scheme of symmetric divisions in Figure 11 is based on the observation that when single cells from myogenic cell cultures are used for establishing clones, one finds that the clones consist of 1, 2, 4, 8, 16, or more than 16 cells. When more than 16 cells are formed, they always consist of both postmitotic and further-dividing cells. On the other hand, when fewer than 16 cells are formed, they are all postmitotic, and occur in numbers corresponding to 2^n (where n = 0, 1, 2, 3, or 4), suggesting that there are 4 obligate divisions after an asymmetric division of the presumptive myoblast. Though the idea of a stem cell population as suggested by Quinn et al.[98] need not be considered as finally established, it explains most of the experimental observations reported so far.

Many workers have observed that when myogenic cells are grown in vitro, they continue to divide. Alignment and fusion are possible only at optimum cell densities. Whether the cells continue to divide indefinitely if passaged during exponential growth has been debated for some time. The existence of established myogenic cell lines suggests that this possibility exists. Hence, as an alternative to the idea of the obligatory four successive symmetric divisions, it has been suggested that fusion-competent myoblasts may result from a lengthening of the G_1 phase of the cell cycle. This lengthening of the G_1 phase may be associated with the onset of the transition to the differentiated state. Myoblasts undergoing this transition repress the synthesis of products characterizing the myoblasts, activate the synthesis of products characteristic of the differentiated muscle, and acquire fusion competence. Whether or not such a cell will fuse with another cell will depend on the probability of its coming in contact with another fusion competent cell and not just any other cell.[100] Though the two alternative theories need to be examined more critically, it appears that the information available at present favors the idea of quantal obligatory divisions.

B. Fusion of Myoblasts

Myotube formation can be observed directly in vitro. However, fusion per se is difficult to demonstrate in tissue culture unequivocally. Since the plasma membrane is not seen by light microscopy, the origin of multinucleate myotubes has to be elucidated by other methods. Several questions may be posed in this connection. Do the myotubes arise by actual fusion of cells or by a process of endomitosis? If fusion does occur, do the nuclei in the myotubes also continue to divide still further? Do all the nuclei retain their functional state?

These questions can now be considered as answered definitively. Myoblasts with labeled nuclei mixed with unlabeled ones fuse to form myotubes in which some nuclei are labeled and the others not, roughly in a proportion similar to the mixture of cells with and without the label. It is also known that the nuclei in the myotubes are postmitotic, i.e., they do not incorporate tritiated thymidine. These facts suggest that the multinucleate myotubes are formed exclusively by the fusion of myoblasts. Additional evidence leading to the same conclusion has come from a study of allophenic mouse embryos. The allophenic embryos are mosaics obtained by introducing a cell from the inner cell mass of a blastocyst into another blastocyst so that the grafted cell and its progeny get incorporated into the developing embryo. Practically all the organs of the allophenic mouse embryo are thus a mosaic of cells from two distinct individuals.

The experiments of Mintz and Baker[101] involved the use of two strains of mice that differed in the isozymes of isocitrate dehydrogenase. The enzyme is a dimer and three isozymes are possible, depending on the types of polypeptides synthesized in the cell. If both types of polypeptides (A and B) are formed within a cell, three isozymes (AA, AB, and BB) will be synthesized in it. On the other hand, if only one type of polypeptide is formed in a cell, only one isozyme (AA or BB) is possible. For a general account of isozymes see Rider and Taylor.[102] In a strain homozygous for A or B, only one type of isozyme is found in all tissues of the body. In a hybrid, all three types will be found since each cell has the genes for the two types of polypeptides. Obviously, this is due to the fact that cells of the two genotypes have fused during fertilization once and for all. In case of allophenic individuals, which are mosaics of cells derived from the two strains, the tissue cells produce either A or B polypeptides. Since the organs are a mixture of the two cell types, the isozymes extracted from them would be of both types, viz., AA and BB. The skeletal muscles of the allophenic mice, however, produce all three isozymes. This is possible because the muscle fibers are syncytia derived from cells originated from both the strains (Figure 12), and both types of polypeptides are formed in the same cytoplasmic milieu.

The evidence outlined above shows clearly that the myotubes are derived from the fusion of myoblasts and that the nuclei in the syncytium are all functionally active.

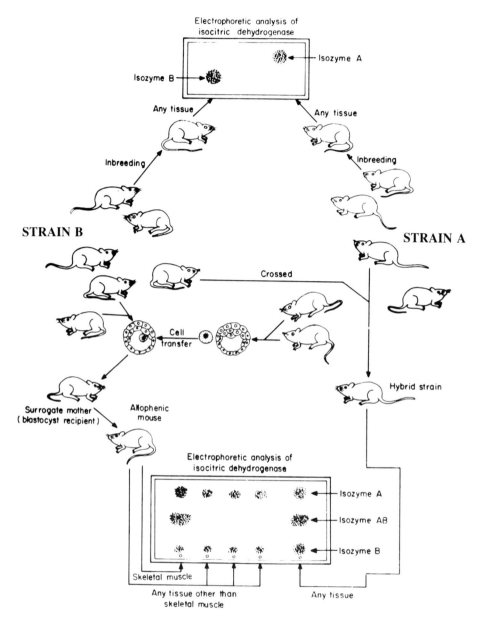

FIGURE 12.   Experimental evidence to show that the skeletal muscle tissue arises from the fusion of cells. See the text. (Adapted from Mintz, B. and Baker, W. W., *Proc. Natl. Acad. Sci. U.S.A.*, 58, 592, 1967.)

1. Fusion-Competent Cells

Cell fusion is a highly specialized developmental process and obviously fusion competence is not possessed by all tissue cells. Identification of the fusion-competent cells thus becomes important. Besides, how the fusion competence is acquired by a certain cell clone during development needs to be investigated. Myoblasts in vitro assume a true spindle shape and thus look different from the typically triradiate fibroblasts, which are also present in cultures of cells obtained from embryonic muscle. Besides, myoblasts when added to cultures of already formed myotubes generally orientate parallel to the latter, whereas fibroblasts do not show any such orientation. Myoblasts can fuse with myoblasts or with already formed myotubes, but not with other cell types. A curious aspect of this specificity in myoblast

Table 3
FORMATION OF HYBRID MUSCLE FIBERS IN CULTURES OF THIGH MUSCLE CELLS OF DIFFERENT GENETIC ORIGIN

Sources of cells[a]

Thymidine ³H-treated cultures	Untreated cultures	Labeled nuclei in fibers of secondary cultures (%)
Rat	—	90 ± 2[b]
Rat	Rabbit	55 ± 12[b]
Rat	—	99 ± 2
Rat	Calf	37 ± 9[b]
Rat	—	93 ± 6[b]
Rat	Chick	51 ± 2.6[b]

[a] In experiments involving cells of different animal origin, the labeled and unlabeled cells were mixed in equal numbers to set up the secondary cultures.

[b] Standard deviations calculated from weighted values according to the number of nuclei in each fiber.

From Yaffe, D. and Feldman, M., *Dev. Biol.*, 11, 300, 1965. With permission.

fusion is that the cells from different species can fuse and form myotubes. Thus the formation of hybrid myotubes from rat myoblasts mixed with rabbit, calf, or chick myoblasts has been demonstrated (Table 3).[103,104]

Carlsson et al.[105] confirmed the formation of hybrid myotubes by visual demonstration of the different nuclei in hybrid myotubes. Fluorescence labeled antibody against nuclear envelope of chick cells was used for the detection of chick nuclei in chick-rat hybrid myotubes. Attempts have been made to see if heterotypic cells fuse with chick myoblasts. Fibroblasts, kidney and liver cells, chondroblasts, and smooth and cardiac muscle cells were each labeled with ³H-thymidine and mixed with myogenic cells. The labeled cells were excluded from the myotubes.[96,97] Even myoblasts do not fuse before they acquire fusion competence. Whether high cell density permitting physical contact of aligned myoblasts is sufficient to make them fusion competent has been debated. There are reports indicating that a macromolecular factor present in conditioned media can cause precocious fusion.[106,107] As described earlier, there is now clear evidence that fusion competence appears as a distinct process in the ontogeny of the myogenic cells. The fusion-competent myoblasts are morphologically distinct and have a different surface topography from the other myoblasts. In this stage, they begin to show terminal differentiation as indicated by the synthesis of muscle specific products. It appears that fusion is not a prerequisite for the initiation of synthetic activity, though the muscle specific substances increase enormously only after fusion.

Myogenic cells do not show the property of contact inhibition. The characteristic prefusion alignment shown by myogenic cells in vivo as well as in vitro does not seem to be an essential prerequisite for fusion. The spindle shape of the cells is lost when microtubules are depolymerized by treatment with colchicine; however, the cells can undergo fusion and form multinucleate syncytia.[108] Knudsen and Horowitz[109] have demonstrated that fusion-competent cells held in suspension also fuse to form spherical multinucleate "myoballs".

2. Monitoring Myoblast Fusion

When myogenic cells are grown in vitro, they multiply and eventually fuse forming multinucleate structures. The progress of fusion is, however, not easy to follow since it is a very slow process. When cells are grown in low (≈ 160 μM) calcium media, they can be withheld from fusion. Fusion competence is, however, acquired in such media. Now if normal calcium levels are restored (≈ 1960 μM) fusion proceeds rapidly. Progress of fusion can be monitored by holding cells in suspension so that myoballs are formed. Another assay to study the kinetics of myoblast fusion has been described.[110] In this procedure, one population of myogenic cells is treated with the lipid fluorescent dye 3-3'-dioctadecylindotricarboxonene and mixed with unlabeled cells. When the cells fuse, the dye will diffuse from one cell into another. It is contended that this assay is more reproducible and reliable.[110]

3. Requirement of Calcium for Fusion

An essential requirement of myoblast fusion is calcium. As mentioned earlier, the prefusion cells can be prevented from fusing by keeping them in a low calcium medium. Other ions (Mg^{2+}, Zn^{2+}, Mn^{2+}, etc.) cannot substitute Ca^{2+} though some fusion occurs[111,112] in the presence of Sr^{2+}. It appears that Ca^{2+} in the external medium is required for both recognition and alignment by the myoblasts. Fusion can also be prevented by media containing calcium chelators at a stage after alignment and before membrane fusion. Trifluoperazine, a calmodulin antagonist, can inhibit myoblast fusion.[113] When Ca^{2+} channels are blocked chemically by the calcium channel blocker, D600, fusion is diminished significantly though alignment is not inhibited. On the other hand, when cultures are treated with the calcium ionophore A 23187, fusion is enhanced.[114,115] There is also a measurable increase in the Ca^{2+} uptake just before membrane fusion occurs. Thus an influx of Ca^{2+} is an essential prerequisite for myoblast fusion. A link between myoblast fusion, and prostaglandin PGE_1 and cyclic AMP has also been suggested.

4. Membrane Alterations and Fusion of Myoblasts

Characterization of cell surface alterations during myogenesis has been attempted by several workers. These studies aim at revealing if any particular surface proteins or glycoproteins appear or disappear as a part of the mechanism underlying the recognition and eventual fusion of the myoblasts. Several different methods of labeling have been employed, since a single method may not label all the surface proteins. For instance, lactoperoxidase-mediated iodination would fail to label the surface proteins that have no tyrosine residues accessible to the enzyme. Walsh and Phillips[116] employed lactoperoxidase-mediated iodination, and periodate oxidation of sialoglycoproteins followed by tritiated borohydrate labeling (see Figure 5, Chapter 1 in Volume I). The labeled proteins or glycoproteins were separated by polyacrylamide electrophoresis and exposed to photographic sensitive material or "stained" with radio-iodinated Con A or WGA as appropriate. By these methods, a number of surface proteins were found to appear concomitant with fusion; other proteins/glycoproteins disappeared from the myoblasts after fusion. Though it is not possible at present to assign any role to these surface proteins in cell adhesion and/or recognition, it is clear that important cell surface changes occur when myotubes are formed.

Inasmuch as the myoblasts have to establish a close contact as preliminary to fusion, one may assume *a priori* that they have to recognize homotypic cells. The recognition is effected presumably through a molecular mechanism at the cell surface, and the most promising candidates for this role would be carbohydrates. In other systems, we have seen that lectins can detect the presence of such molecules. Lectin-mediated inhibition of myoblast fusion has been demonstrated (Table 4; for details see Den et al.[117]).

Subsequently, other studies have led to the discovery of a muscle-specific lectin, which can be obtained from the pectoral muscle of 16-day chick embryos.[118] Media containing lactose can extract the lectin under mild conditions (Figure 13). The lectin thus obtained (after dialysis to remove lactose) can cause agglutination of glutaraldehyde-treated rabbit erythrocytes. The lectin-mediated agglutination can be inhibited specifically by lactose.

Table 4
INHIBITION OF MYOBLAST FUSION BY LECTINS[a]

	Percentage nuclei in myotubes		Cells per field	
Lectin	Expt. 1	Expt. 2	Expt. 1	Expt. 2
None	50	59	165	290
Con A	25 (50)[b]	32 (54)	100 (61)	150 (52)
Lens culinaris lectin	24 (48)	37 (63)	140 (85)	270 (93)

[a] Cultures were treated continuously with 30 μg/mℓ of each lectin from 15 hr after plating until sacrifice at 50 hr.
[b] Numbers in parentheses are percentage of control values.

From Bischoff, R., *Cell Surface Rev.*, 5, 153, 1978. With permission.

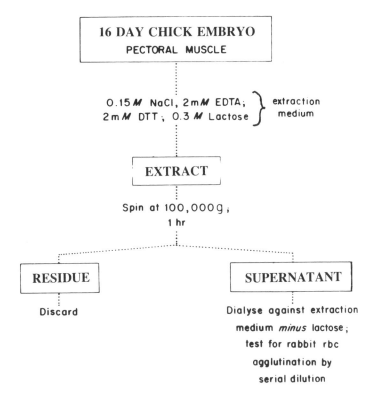

FIGURE 13.   The procedure followed in the extraction of chick muscle lectin. (Adapted from Novak, T. P., Haywood, P. L., and Barondes, S. H., *Biochem. Biophys. Res. Commun.*, 68, 650, 1976.)

Dithiodigalactoside is also effective in competitive inhibition of rabbit erythrocyte agglutination in the presence of muscle lectin. Nowak et al.[118] have tested 16 other saccharides including galactose which are much less effective. The precise role of the muscle lectin in myogenesis is, however, not clear. Den and Chin[119] indicate that the muscle lectin may not be involved in myotube formation. Obviously more needs to be known to show clearly if myoblast recognition during myogenesis is effected through muscle-specific protein(s) with carbohydrate specificity. Another candidate for a role in myoblast recognition and adhesion could be fibronectin. Experimental evidence, however, indicates that no such role is played by fibronectin.[120,121]

Myogenic cells in vitro release into the culture medium a glycoprotein complex called adheron, which promotes adhesion of myoblasts. The adhesion-mediating activity was found to have a sedimentation value of 16 S in sucrose density gradients in the absence of Ca^{2+}. The adheron was shown to consist of a glycosaminoglycan and several proteins, including collagen and fibronectin.[122] The myoblast cell line L6 and an adhesion deficient variant of L6 (designated M 3A) released adherons into the medium. The adheron from M 3A differed from that of L6 in several properties. Compared with the adheron of L6 cells, that of M 3A cells was much less effective in promoting mutual adhesion of normal myoblasts and their adhesion to the substratum. The two adherons also differed in their sedimentation velocities in sucrose gradients. Further, the M 3A adheron lacked chondroitin sulfate and was deficient in collagen and fibronectin content. Schubert and LaCorbiere,[123] who revealed these interesting differences, suggest that the adherons bind mutually and also on the myoblasts. Mutual binding of adherons is calcium-dependent, unlike their binding on myoblasts. Adhesion of myoblasts seems to be brought about by the adherons acting as links between cells. The adherons of the M 3A variant are unable to bind mutually, and hence the myoblasts fail to adhere. These studies indicate that the adhesion of myoblasts prior to fusion is controlled by the myoblasts themselves, and a role for the connective tissue cells is not evident.

Establishment of cytoplasmic continuity during fusion of cells is possible only if the intervening lipid layers are disrupted. Any agent that increases membrane fluidity should therefore enhance myoblast fusion, and by the same token, the agents that confer lipid membrane rigidity should prevent fusion. It has been demonstrated that an increased fluidity of the membrane precedes fusion. Adding steric acid that reduces membrane fluidity delays fusion, whereas oleic acid that increases fluidity is found to enhance fusion.[124] In general, it appears that membrane destabilization, which is necessary for the establishment of cytoplasmic continuity during myoblast fusion, does not occur over a large area of the cells in contact. Membrane breakdown appears to be restricted to a small area initially, and this is followed by withdrawal of the membrane over the other region. Fulton et al.[125] have described an interesting method of revealing the surface changes in prefusion and fused myoblasts. The cells are extracted with a detergent that removes the membrane lipids, leaving behind the *surface lamina*, i.e., the surface framework consisting of proteins/glycoproteins. Scanning and transmission electron microscopy of these lipid-extracted cells reveals interesting structural details. Using this method, it has been demonstrated that myoblasts preparing to fuse show numerous lacunae in the surface lamina coinciding with the domains of the cell surface where Con A binding sites are absent. These lacunae are not caused by the extraction procedure; they correspond to the regions where glycoproteins are absent. The lacunae appear to be related to the fusion of myoblasts, disappearing in multinucleate myotubes. It has been suggested by Fulton et al.[125] that the lacunae in the surface lamina relate to the requirements of cell fusion. The protein-free regions are probably the sites where the fusing plasma membranes come close and, on destabilization of lipid layers, a cytoplasmic continuity is established.

An independent line of evidence corroborates the idea that fusion is initiated at a lipid-rich and protein-free domain of the plasma membrane. According to a model presented by Kalderon and Gilula,[126] the earliest event in myoblast fusion is the formation of unilamellar, particle-free vesicles in the cytoplasm. (In freeze-fracture parlance, ''particle free'' refers to the lipid membrane free of proteins spanning its thickness.) An interaction of the vesicle and the plasma membrane results in a destabilization of the latter, presumably by rendering it enriched in phospholipids. Subsequently the vesicle membrane fuses with the plasma membrane in the domain of interaction. Two plasma membrane particle-free regions of adjacent myoblasts can then fuse to form a single bilayer and thus establish cytoplasmic continuity in a small region, about 15 Å in diameter. Then the cytoplasmic bridge enlarges.

It has been suggested that a limited proteolytic action on the cell surface could facilitate fusion by reducing steric or charge restraints and thereby bringing about a close apposition of the plasma membranes of the fusing myoblasts. The requirement of a metalloendoprotease activity has been demonstrated in the fusion of rat myoblasts.[127] The metalloprotease inhibitor, 1-10-phenanthroline, prevents the fusion of myoblasts at an effective concentration of 30 μg/mℓ. Besides, competitive inhibition of the enzyme activity by the addition of synthetic substrates (carbobenzoxy-ser-leu-amide and carbobenzoxy-tyr-phe-amide) can prevent myoblast fusion in cultures.

The foregoing account indicates clearly that at least three steps are involved in the fusion of myoblasts. First, two fusion-competent myoblasts must come together. They must then adhere to each other. Finally the adhesion should become sufficiently close to aid eventual fusion. Formation of protein/glycoprotein-free domain may diminish the repulsive forces between the two cells. Admittedly the biophysical feasibility of these events needs to be examined critically. We can, however, conclude that some general aspects of the mechanism of myoblast fusion have been clarified by the recent research.

The "muscles" recognized by anatomists are composed of the muscle fibers. Large numbers of muscle fibers (including the accompanying satellite cells) are held in parallel bundles constituting the structures called fascicles. The latter are organized into the "muscles" individualized by the presence of a connective tissue envelope (epimysium) and the presence of a tendon at one or both ends. The anatomy of the muscular system is highly conserved in vertebrate evolution, thus permitting the recognition of homologies of different muscles. During ontogeny, the muscles, including their normal association with the skeleton, develop harmoniously into functional units which possess amazing abilities of coordinated function. Certainly, the laying down of myotubes and their association with the skeletal system during embryonic development must be a precisely controlled process. We have mentioned earlier that even the muscle fibers located in distal parts such as the extremities of limbs originate from the somties, though the connective tissue investments (epimysium, perimysium, and endomysium) and tendons are derived locally from the connective tissue of the somatopleure. Given that there is a correlation between a somite and a group of limb muscles,[91] how is it that the myogenic cells from the myotome reach the correct anatomical sites? Further, what forces hold them aligned in parallel arrays so that they assume their characteristic anatomical features after fusion? What determines the association of tendons with the fascicles leading to the development of muscles described as uni-, bi-, and multipennate?

Experimental embryologists have shown that heterotopically transplanted somites give rise to myogenic tissues, and eventually the muscles develop appropriate to the new anatomical relations. Evidently, the cells of the myotome have no intrinsic program of a developmental pattern. It therefore emerges from their interaction with the surrounding connective tissue cells and the extracellular matrix. Such a role for the extracellular matrix has been demonstrated by Bayne et al.[107] Extracellular proteins related to the development of myotendinous structures have been detected recently.[128,129] Using an in vitro test system, Venkatasubramanian and Solursh[130] have shown that a gradient of platelet-derived growth factor can attract myoblasts chemotactically. It is also probable that the supramolecular organization of fibronectin, collagen, and other material in the developing limb can align the cells and group them in characteristic ways. The possible mechanisms of the division of muscle masses and the role of extracellular matrix components, especially fibronectin and glycosaminoglycans, has been investigated.[131] Studies on vertebrate limb development have shown clearly that an ectodermal ridge at the apex of the limb rudiment influences the underlying cells and promotes the elongation of the developing limb.[132,133] How the ectodermal ridge controls the organization of the underlying cells is far from clear. Certainly, it is not as mysterious as it looked once. The information already available sharpens the

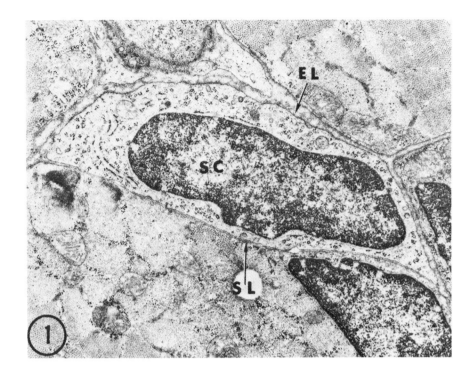

FIGURE 14.   Electron micrograph of a satellite cell (SC) between external lamina (EL) and sarcolemma
(SL) of an uninjured myofiber in the soleus muscle of a rat. (From Snow, M. H., in *Muscle Regeneration,*
Mauro, A., Ed., Raven Press, New York, 1979, 92. With permission.)

focus for future attention. The tissues possessing "inducing" influences are evidently or-
dering the connective tissue cells and the extracellular matrix with which the myogenic cells
interact.

C. Myoblast Fusion During Muscle Regeneration, Hypertrophy, and Neoplasia

The skeletal muscle is capable of regeneration following injury. Experimental injury may
be caused by a variety of methods such as mincing the muscle or exposing it to very low
temperatures, as in frostbite. Adult muscles undergo atrophy following denervation. Muscle
injury may result from other causes also. The extra-ocular muscle fibers degenerate when
albino rats are exposed continuously to high intensity of incandescent or fluorescent light.
Following the injury, the tissue can regenerate if suitable conditions exist. Regeneration can
occur in dimensions ranging from the repair of a single muscle fiber or a small group of
fibers, to the formation of an entire muscle from minced fragments.

The identity of the cells giving rise to the regenerated muscle has been debated for some
time. There are two contending hypotheses regarding the precursor cells that give rise to
the regenerated muscle. According to some investigators, a plasmalemma can be formed
around the nuclei of injured muscle fibers, and the cells formed in this manner constitute
the precursors of the regenerate. It is proposed that these cells divide, eventually fusing to
form multinucleate muscle fibers.[134,135] The alternative hypothesis maintains that reserve
mononucleated cells in the muscle fibers proliferate after injury and eventually fuse to form
new myotubes. The reserve cells, commonly called satellite cells, were first described by
Mauro.[136] The satellite cells are fusiform mononucleated cells situated beneath the external
lamina of the muscle fiber (Figure 14). According to this hypothesis, the satellite cells are
liberated from their association with the muscle fiber when the latter is injured. It is proposed
that the satellite cells thus released, multiply and eventually fuse forming syncytial myotubes.

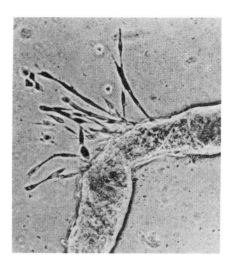

FIGURE 15.   Outgrowth of myogenic cells from a muscle fiber explanted in vitro. (From Bischoff, R., *Anat. Rec.*, 182, 215, 1975. With permission.)

The evidence implicating the satellite cells in the process of muscle regeneration is much more convincing compared with the support to the alternative hypothesis. Proliferation of the satellite cells seems to be stimulated by a soluble factor present in the extracts of young skeletal muscle. Extracts of older muscle promote greater growth of connective tissue rather than the descendants of satellite cells.[137] These observations accord with the generally observed fact that younger animals have greater powers of regenerating injured muscle. The failure of muscle regeneration seems to be associated with the rapid growth of the connective tissue after injury, and laying down collagen and other extra-cellular matrix material as a scar tissue.

In vitro studies have revealed that adult muscle fibers can liberate mononucleate cells when treated with trypsin or pronase. Electron microscopic examination of the muscle fibers after the treatment with these enzymes reveals that the satellite cells enclosed within the basal lamina are absent. The liberated satellite cells multiply and form typical myotubes, which can show spontaneous contraction (Figure 15). Some other proteolytic enzymes (papain, collagenase, ficin) also liberate viable mononucleated cells from adult muscle fibers. These cells are fibroblastic and do not form myofibrils. When muscle fibers treated with these enzymes are examined by electron microscopy, the satellite cells are found to be intact, enclosed within the basal lamina of the fiber. These facts constitute sound evidence to implicate the satellite cells as the only cellular source of skeletal muscle regeneration. For a detailed discussion on the subject, see Bischoff,[138,139] and Snow.[140]

Muscular hypertrophy (increase in muscle mass as a result of exercise) also involves the formation of new fibers. The satellite cells have been implicated in this process also.[141] Rhabdomyosarcoma, a tumor derived from the skeletal muscle, also seems to arise from the satellite cells.[142]

REFERENCES

1. **Brachet, J.,** A comparison of nucleocytoplasmic interaction in *Acetabularia* and in eggs, in *Progress in Developmental Biology,* Saur, H. W., Ed., Gustav Fischer Verlag, Stuttgart, 1981, 15.

2. **Gurdon, J. B. and Woodland, H. R.,** The cytoplasmic control of nuclear activity in animal development, *Biol. Rev.,* 43, 233, 1968.

3. **Cocking, E. C.,** Plant-animal cell fusions, in *Cell Fusion, Ciba Foundation Symp.,* 103, Pitman Books, London, 1984, 119.

4. **Lucy, J. A.,** Mechanisms of chemically induced cell fusion, in *Cell Surface Rev. 5,* 1978, chap. 6.

5. **Poste, G. and Pasternak, C. A.,** Virus-induced cell fusion, in *Cell Surface Rev. 5,* 1978, chap. 7.

6. **Knutton, S.,** Studies on membrane fusion. VI. Mechanism of the membrane fusion and cell swelling stages of Sendai virus-mediated cell fusion, *J. Cell Sci.,* 43, 103, 1980.

7. **Parsegian, V. A., Rand, R. P., and Gingell, D.,** Lessons for the study of membrane fusion from membrane interactions in phospholipid system, in *Cell Fusion, Ciba Foundation Symp.,* 103, Pitman Books, London, 1984, 9.

8. **Zimmerman, U. and Vienken, J.,** Electric field-induced cell-to-cell fusion, *J. Membrane Biol.,* 67, 165, 1982.

9. **Evans, D. A., Bravo, J. E., and Gleba, Y. Y.,** Somatic hybridization — fusion methods, recovery of hybrids and genetic analysis, *Int. Rev. Cytol.,* 16 (Suppl.), 143, 1983.

10. **Kohler, G. and Milstein, C.,** Continuous cultures of fused cells secreting antibody of predetermined specificity, *Nature (London),* 256, 495, 1975.

11. **Furusawa, M.,** Cellular microinjection by cell fusion: technique and applications in biology and medicine, *Int. Rev. Cytol.,* 62, 29, 1980.

12. **Fischman, D. A. and Hay, E. D.,** Origin of osteoclasts from mononuclear leucocytes in regenerating newt limbs, *Anat. Rec.,* 143, 329, 1962.

13. **Jotereau, F. V. and LeDouarin, N. M.,** The developmental relationship between osteocytes and osteoclasts: a study using the quail-chick nuclear marker in endochondral ossification, *Dev. Biol.,* 63, 253, 1978.

14. **Enders, A. C.,** Formation of the syncytium from cytotrophoblast in the human placenta, *Obstet. Gynecol.,* 25, 378, 1965.

15. **Midgley, A. R., Jr., Pierce, G. B. R., Deneau, G. A., and Gosling, J. R. G.,** Morphogenesis of syncytiotrophoblast *in vitro*: an autoradiographic demonstration, *Science,* 141, 349, 1963.

16. **Nagano, T. and Suzuki, F.,** Cell junctions in the seminiferous tubule and the excurrent duct of the testis. Freeze fracture studies. *Int. Rev. Cytol.,* 81, 163, 1983.

17. **O'Rand, M. G. and Romrell, L. J.,** Appearance of regional surface autoantigens during spermatogenesis: comparison of anti-testis and anti-sperm antisera, *Dev. Biol.,* 75, 431, 1980.

18. **Koehler, J. K.,** The mammalian sperm surface: studies with specific labelling techniques, *Int. Rev. Cytol.,* 51, 73, 1978.

19. **Holt, W. V.,** Membrane heterogeneity in the mammalian spermatozoon, *Int. Rev. Cytol.,* 87, 159, 1984.

20. **Gordon, M. and Dandekar, P. V.,** Fine structural localization of phosphatase activity on the plasma membrane of the rabbit sperm head, *J. Reprod. Fertil.,* 36, 211, 1977.

21. **Lopo, A. and Vacquier, V. D.,** Sperm-specific surface antigenicity common to seven animal phyla, *Nature (London),* 288, 397, 1980.

22. **Nicolson, G. L. and Yanagimachi, R.,** Mobility and the restriction of mobility of plasma membrane lectin-binding components, *Science,* 184, 1294, 1974.

23. **Virtanen, I., Badley, R. A., Paasivuo, and Lehto, V-P.,** Distinct cytoskeletal domains revealed in sperm cells, *J. Cell Biol.,* 99, 1083, 1984.

24. **Arora, R., Dinakar, N., and Prasad, M. R. N.,** Biochemical changes in the spermatozoa and luminal contents of different regions of the epididymis of the rhesus monkey, *Macaca mulatta, Contraception,* 11, 689, 1975.

25. **Rajalaksmi, M., Arora, R., Bose, T. K., Dinakar, N., Gupta, G., Thampan, T. N. R. V., Prasad, M. R. N., Anand Kumar, T. C., and Moudgal, N. R.,** Physiology of the epididymis and induction of functional sterility in the male, *J. Reprod. Fertil.,* 24(Suppl.), 71, 1976.

26. **Fenderson, B. A., O'Brien, D. A., Millette, C. F., and Eddy, E. M.,** Stage-specific expression of three cell surface carbohydrate antigens during murine spermatogenesis detected with monoclonal antibodies, *Dev. Biol.,* 103, 117, 1984.

27. **Brown, C. R., Von Glos, K. I., and Jones, R.,** Changes in plasma membrane glycoproteins of rat spermatozoa during maturation in the epididymis, *J. Cell Biol.,* 96, 256, 1983.

28. **Jones, R., Brown, C. R., Von Glos, K. I., and Gaunt, S. J.,** Development of a maturation antigen on the plasma membrane of rat spermatozoa in the epididymis and its fate during fertilization, *Exp. Cell Res.,* 156, 31, 1985.

29. **Austin, C. R.,** Observations on the penetration of the sperm into the mammalian egg, *Aust. J. Sci. Res.,* B 4, 581, 1951.

30. **Chang, M. C.,** Fertilizing capacity of spermatozoa deposited into the fallopian tube, *Nature (London),* 168, 697, 1951.

31. **Farooqui, A. A.,** Biochemistry of sperm capacitation, *Int. J. Biochem.,* 15, 463, 1983.

32. **Bedford, J. M.,** Significance of the need for sperm capacitation before fertilization in eutherian mammals, *Biol. Reprod.,* 28, 108, 1983.
33. **Gordon, M., Dandekar, P. V., and Bartoszewicz, W.,** Ultrastructural localization of surface receptors, *J. Reprod. Fertil.,* 36, 211, 1974.
34. **Gordon, M., Dandekar, P. V., and Bartoszewicz, W.,** The surface coat of epididymal ejaculated and capacitated sperm, *J. Ultrastruct. Res.,* 50, 199, 1975.
35. **Ward, C. R. and Storey, B. T.,** Determination of the time course of capacitation in mouse spermatozoa using a chlorotetracycline fluorescence assay, *Dev. Biol.,* 104, 287, 1984.
36. **Shapiro, B. M. and Eddy, E. M.,** When sperm meets egg: biochemical mechanisms of gamete interaction, *Int. Rev. Cytol.,* 66, 257, 1980.
37. **Dan, J. C.,** Studies on acrosome. III. Effect of calcium deficiency, *Biol. Bull.,* 107, 335, 1954.
38. **Tilney, L. G., Kiehart, D. P., Sardet, C. and Tilney, M.,** Polymerization of actin. IV. Role of Ca^{2+} and H^+ in the assembly of actin and in membrane fusion in the acrosomal reaction of echinoderm sperm, *J. Cell Biol.,* 77, 536, 1978.
39. **Schackman, R. W., Eddy, E. M., and Shapiro, B. M.,** The acrosome reaction of *Strongylocentrotus purpuratus* sperm. Ion requirements and movements, *Dev. Biol.,* 65, 483, 1978.
40. **Smith, M., Peterson, R. N., and Russel, L. D.,** Penetration of zona-free hamster eggs by boar sperm treated with the ionophore A23187 and inhibition of penetration by anti-plasma membrane antibodies, *J. Exp. Zool.,* 225, 157, 1983.
41. **Tilney, L. G. and Kallenbach, N.,** Polymerization of actin. VI. The polarity in the acrosomal process and how it might be determined, *J. Cell Biol.,* 81, 608, 1979.
42. **Tilney, L. G. and Inoue, T.,** Acrosomal reaction of *Thyone* sperm. II. The kinetics and possible mechanism of acrosomal process elongation, *J. Cell Biol.,* 93, 820, 1982.
43. **Epel, D. and Vacquier, V. D.,** Membrane fusion events during invertebrate fertilization, in *Cell Surface Rev.,* 5, 1978, chap. 1.
44. **Tegner, M. J. and Epel, D.,** Scanning electron microscope studies of sea urchin fertilization. I. Eggs with vitelline layers, *J. Exp. Zool.,* 197, 31, 1976.
45. **Monroy, A. and Baccetti, B.,** Morphological changes of the surface of the egg of *Xenopus laevis* in the course of development. I. Fertilization and early cleavage, *J. Ultrastruct. Res.,* 50, 131, 1975.
46. **Vacquier, V. D.,** Isolation of intact cortical granules from sea urchin eggs: calcium ions trigger discharge, *Dev. Biol.,* 43, 62, 1975.
47. **Eager, D. D., Johnson, M. H., and Thurley, K. W.,** Ultrastructural studies on the surface membrane of the mouse egg, *J. Cell Sci.,* 22, 345, 1976.
48. **Yanagimachi, R. and Nicolson, G. L.,** Lectin binding properties of hamster egg zona pellucida and plasma membrane during maturation and pre-implantation development, *Exp. Cell Res.,* 100, 249, 1976.
49. **Nicolson, G. L., Yanagimachi, R., and Yanagimachi, H.,** Ultrastructural localization of lectin binding sites on the zonae pellucidae and plasma membranes of mammalian eggs, *J. Cell Biol.,* 66, 263, 1975.
50. **Saling, P. M., Rainer, L. M., and O'Rand, M. G.,** Monoclonal antibody against mouse sperm blocks a specific event in the fertilization process, *J. Exp. Zool.,* 227, 481, 1983.
51. **Bedford, J. M.,** Why mammalian gametes don't mix, *Nature (London),* 291, 286, 1981.
52. **Glabe, C. G., Grabel, L. A., Vacquier, V. D., and Rosen, S. D.,** Carbohydrate specificity of sea urchin sperm bindin: a cell surface lectin mediating sperm-egg adhesion, *J. Cell Biol.,* 94, 123, 1982.
53. **Vacquier, V. D. and Moy, G. W.,** Isolation of bindin: the protein responsible for adhesion of sperm to sea urchin eggs, *Proc. Natl. Acad. Sci. U.S.A.,* 74, 2456, 1977.
54. **Vacquier, V. D.,** Purification of sea urchin bindin by DEAE-cellulose chromatography, *Anal. Biochem.,* 129, 497, 1983.
55. **Vacquier, V. D.,** Rapid immunoassays for the acrosome reaction of sea urchin sperm utilizing antibody to bindin, *Exp. Cell Res.,* 153, 281, 1984.
56. **Schmell, E., Earls, B. J., Breaux, C., and Lennarz, W. J.,** Identification of a sperm receptor on the surface of the eggs of the sea urchin *Arbacia punctulata, J. Cell Biol.,* 72, 35, 1977.
57. **Ahuja, K. K.,** Fertilization studies in the hamster. The role of cell surface carbohydrates, *Exp. Cell Res.,* 140, 353, 1982.
58. **Huang, T. T. F. and Yanagimachi, R.,** Fucoidin inhibits attachment of guinea pig spermatozoa to the zona pellucida through binding to the inner acrosomal membrane and equatorial domains, *Exp. Cell Res.,* 153, 363, 1984.
59. **Rossignol, D. P., Earles, B. J., Decker, G. L., and Lennarz, W. J.,** Characterization of the sperm receptor on the surface of eggs of *Strongylocentrotus purpuratus, Dev. Biol.,* 104, 368, 1984.
60. **Brachet, A.,** Etude sur les localisations germinales et leur potentialité réelle dans l'oeuf parthénogénétique de *Rana fusca, Arch. Biol.,* 26, 337, 1911.
61. **Elnison, R. P. and Manes, M. E.,** Morphology of the site of sperm entry on the frog egg, *Dev. Biol.,* 63, 67, 1978.
62. **Schatten, G.,** Motility during fertilization, *Int. Rev. Cytol.,* 79, 60, 1982.

63. **Bedford, J. M. and Cooper, G. W.,** Membrane fusion events in the fertilization of vertebrate eggs, in *Cell Surface Rev.,* 5, 1978, chap. 2.

64. **Conway, A. F. and Metz, C. B.,** Phospholipase activity of sea urchin sperm: its possible involvement in membrane fusion, *J. Exp. Zool.,* 198, 39, 1976.

65. **Jaffe, L. A.,** Fast block to polyspermy in sea urchin eggs is electrically mediated, *Nature (London),* 261, 68, 1976.

66. **Whitaker, M. J. and Steinhardt, R. A.,** Evidence in support of the hypothesis of an electrically mediated fast block to polyspermy in sea urchin eggs, *Dev. Biol.,* 95, 244, 1983.

67. **Nuccetelli, R. and Grey, R. D.,** Controversy over the fast, partial, temporary block to polyspermy in sea urchins. A revaluation, *Dev. Biol.,* 103, 1, 1984.

68. **Eddy, E. M. and Shapiro, B. M.,** Changes in the topography of the sea urchin egg after fertilization, *J. Cell Biol.,* 71, 35, 1976.

69. **Ancel, P. and Vintemberger, P.,** Récherches sur le déterminisme de la symmétrie bilatérale dans l'oeuf de Amphibiens, *Bull. Biol. Fr. Belg.,* 31, 1, 1948.

70. **Jaffe, L. A., Sharp, A. P., and Wolf, P. D.,** Absence of an electrical polyspermy block in the mouse, *Dev. Biol.,* 96, 317, 1983.

71. **Steinhardt, R. A., Epel, D., Carroll, E. J., and Yanagimachi, R.,** Is Ca^{2+} ionophore a universal activator of unfertilized eggs?, *Nature (London),* 252, 41, 1974.

72. **Ridgeway, E. B., Gilkey, C. J., and Jaffe, L. F.,** Free calcium increases explosively in activating medaka eggs, *Proc. Natl. Acad. Sci. U.S.A.,* 74, 623, 1977.

73. **Vacquier, V. D.,** Dynamic changes of the egg cortex, *Dev. Biol.,* 84, 1, 1981.

74. **Gabel, C. A., Eddy, E. M., and Shapiro, B. M.,** After fertilization sperm-specific components remain as a patch in sea urchin and mouse embryos, *Cell,* 18, 207, 1979.

75. **Gundersen, G. G., Gabel, C. A., and Shapiro, B. M.,** An intermediate state of fertilization involved in internalization of sperm components, *Dev. Biol.,* 93, 59, 1982.

76. **Gaunt, S. J.,** Spreading of sperm surface antigen within the plasma membrane of the egg after fertilization in the rat, *J. Embryol. Exp. Morphol.,* 75, 259, 1983.

77. **Gundersen, G. G. and Shapiro, B. M.,** Sperm surface proteins persist after fertilization, *J. Cell Biol.,* 99, 1343, 1984.

78. **Knox, R. B., Clarke, A., Harrison, S., Smith, P., and Marchalonis, J. J.,** Cell recognition in plants: determination of the stigma surface and their pollen interactions, *Proc. Natl. Acad. Sci. U.S.A.,* 73, 2788, 1976.

79. **Shivanna, K. R.,** Recognition and rejection phenomena during pollen-pistil interaction, *Proc. Indian Acad. Sci.,* 88B, 115, 1978.

80. **Heslop-Harrison, J.,** Pollen-stigma interaction and cross-incompatibility in grasses, *Science,* 215, 1358, 1982.

81. **Shivanna, K. R.,** Pollen-pistil interaction and control of fertilization, in *Experimental Embryology of Vascular Plants,* Johri, B. M., Ed., Springer-Verlag, Berlin, 1982, chap. 7.

82. **Raff, J. and Knox, R. B.,** Self incompatibility in sweet cherry *Prunus avium, Incompatibility Newsl.,* 8, 36, 1977.

83. **Rangaswamy, N. S.,** Application of *in vitro* pollination and *in vitro* fertilization, in *Applied and Fundamental Aspects of Plant Cell, Tissue and Organ Culture,* Reinert, J. and Bajaj, Y. P. S., Eds., Springer-Verlag, Berlin, 1977, 412.

84. **Roberts, I. N., Stead, A. D., Ockendon, D. J., and Dickinson, H. G.,** Pollen-stigma interactions in *Brassica oleracea* (Review), *Theoret. Appl. Genet.,* 58, 241, 1980.

85. **Roberts, I. A., Harrod, G., and Dickinson, H. G.,** Pollen-stigma interactions in *Brassica oleracea.* I. Ultrastructure and physiology of the stigmatic papillar cells, *J. Cell Sci.,* 66, 241, 1984.

86. **Roberts, I. A., Harrod, G., and Dickinson, H. G.,** Pollen-stigma interactions in *Brassica oleracea.* II. Fate of stigma surface proteins following pollination and their role in the self-incompatibility response, *J. Cell Sci.,* 66, 255, 1984.

87. **Dumas, C., Knox, R. B., and Gaude, T.,** Pollen-pistil recognition: new concepts from electron microscopy and cytochemistry, *Int. Rev. Cytol.,* 90, 239, 1984.

88. **Shivanna, K. R. and Johri, B. M.,** *The Angiosperm Pollen: Structure and Function,* Wiley Eastern, New Delhi, 1985.

89. **Yaffe, D.,** Rat Skeletal muscle cells, in *Tissue Culture: Methods and Applications,* Krause, P. F. and Patterson, M. K., Jr., Eds., Academic Press, New York, 1973, chap. 16.

90. **Chevallier, A., Kiney, M., Mauger, A., and Sengel, P.,** Developmental fate of the somitic mesoderm in the chick embryo, in *Vertebrate Limb and Somite Morphogenesis,* Ede, D. A., Hinchliffe, J. R., and Balls, M., Eds., Cambridge University Press, London, 1977, 421.

91. **Beresford, B.,** Brachial muscles in the chick embryo: the fate of individual somites, *J. Embryol. Exp. Morphol.,* 77, 99, 1983.

92. **Gumpel-Pinot, M., Ede, D. A., and Flint, O. P.,** Myogenic cell movement in the developing avian limb bud in the presence and absence of apical ectodermal ridge (AER), *J. Embryol. Exp. Morphol.,* 80, 105, 1984.

93. **Ede, D. A., Gumpel-Pinot, M., and Flint, O. P.,** Oriented movement of myogenic cells in the avian limb bud and its dependence on presence of the apical ectodermal ridge, *Prog. Clin. Biol. Res.,* 151, 427, 438.

94. **Abbot, J., Schlitz, J., Dienstman, S., and Holtzer, J.,** The phenotypic complexity of myogenic clones, *Proc. Natl. Acad. Sci. U.S.A.,* 71, 1506, 1974.

95. **Sasse, J., Horwitz, A., Pacifici, M., and Holtzer, H.,** Separation of precursor myogenic and chondrogenic cells in early limb bud mesenchyme by a monoclonal antibody, *J. Cell Biol.,* 99, 1856, 1984.

96. **Holtzer, H. and Bischoff, R.,** Mitosis and myogenesis, in *The Physiology and Biochemistry of Muscle as a Food,* Briskey, E. J., Kassen, R. G., and Marsh, B. B., Eds., University of Wisconsin Press, Madison, 1970, 29.

97. **Okazaki, K. and Holtzer, H.,** An analysis of myogenesis *in vitro* using fluorescein-labelled antimyosin, *J. Histochem. Cytochem.,* 13, 726, 1965.

98. **Quinn, L. S., Nameroff, M., and Holtzer, H.,** Age-dependent changes in myogenic precursor cell compartment sizes. Evidence for the existence of a stem cell, *Exp. Cell Res.,* 154, 65, 1984.

99. **Quinn, L. S., Holtzer, H., and Nameroff, M.,** Generation of chick skeletal muscle cells in groups of 16 from stem cells, *Nature (London),* 313, 692, 1985.

100. **Lee, H. U., Kaufman, S. J., and Coleman, J. R.,** Expression of myoblast and myocyte antigens in relation to differentiation and the cell cycle, *Exp. Cell Res.,* 152, 331, 1984.

101. **Mintz, B. and Baker, W. W.,** Normal mammalian muscle differentiation and gene control of isocitrate dehydrogenase synthesis, *Proc. Natl. Acad. Sci. U.S.A.,* 58, 592, 1967.

102. **Rider, C. C. and Taylor, C. B.,** *Isozymes,* Chapman & Hall, London, 1980.

103. **Yaffe, D. and Feldman, M.,** The formation of hybrid multinucleated muscle fibres from myoblasts of different genetic origin, *Dev. Biol.,* 11, 300, 1965.

104. **Maslow, D. E.,** Cell specificity in the formation of multinucleated striated muscle, *Exp. Cell Res.,* 54, 381, 1969.

105. **Carlsson, S. A., Luger, O., Ringertz, N. R., and Savage, R. E.,** Phenotypic expression in erythrocyte × rat myoblast hybrids and in chick myoblast × rat myoblast hybrids, *Exp. Cell Res.,* 84, 47, 1974.

106. **Doering, J. L. and Fischman, D. A.,** A fusion-promoting macromolecular factor in muscle conditioned medium, *Exp. Cell Res.,* 107, 355, 1977.

107. **Bayne, E. K., Anderson, M. J. and Fambrough, D. M.,** Extracellular matrix organization in developing muscle: correlation with acetyl choline receptor aggregates, *J. Cell Biol.,* 99, 1486, 1984.

108. **Fukuda, J., Henkart, M. P., Fischman, G. D., and Smith, T. G., Jr.,** Physiological and structural properties of colchicine-treated chick skeletal muscle cells grown in tissue culture, *Dev. Biol.,* 49, 395, 1976.

109. **Knudsen, K. A. and Horowitz, A. F.,** Tandem events in myoblast fusion, *Dev. Biol.,* 58, 328, 1977.

110. **Neff, N., Decker, C., and Horwitz, A.,** The kinetics of myoblast fusion, *Exp. Cell Res.,* 153, 25, 1984.

111. **Schudt, C. and Pette, D.,** Influence of ionophore A23187 on myogenic cell fusion, *FEBS Lett.,* 59, 36, 1975.

112. **Schudt, C., van der Bosch, J., and Pette, D.,** Inhibition of muscle cell fusion *in vitro* by Mg^{2+} and K^+ ions, *FEBS Lett.,* 32, 296, 1973.

113. **Bar-Sagi, D. and Prives, J.,** Trifluoperazine, a calmodulin antagonist, inhibits muscle cell fusion, *J. Cell Biol.,* 97, 1375, 1983.

114. **David, J. D., See, W. M., and Higginbotham, C-A.,** Fusion of chick embryo skeletal myoblasts: role of calcium influx preceding membrane union, *Dev. Biol.,* 82, 297, 1981.

115. **David, J. D. and Higginbotham, C. A.,** Fusion of chick embryo skeletal myoblasts: interaction of prostaglandin E_1, adenosine 3′: 5′ monophosphate and calcium influx, *Dev. Biol.,* 82, 308, 1981.

116. **Walsh, F. S. and Phillips, E.,** Specific changes in cellular glycoproteins and surface proteins during myogenesis in clonal muscle cells, *Dev. Biol.,* 81, 229, 1981.

117. **Den, H., Malinzak, D. A., Keating, H. J., and Rosenberg, A.,** Influence of concanavalin A, wheat germ agglutinin and soybean agglutinin on the fusion of myoblasts *in vitro, J. Cell Biol.,* 67, 826, 1975.

118. **Nowak, T. P., Haywood, P. L., and Barondes, S. H.,** Developmentally regulated lectin in embryonic chick muscle and a myogenic cell line, *Biochem. Biophys. Res. Commun.,* 68, 650, 1976.

119. **Den, H. and Chin, J. H.,** Endogenous lectin from chick embryo skeletal muscle is not involved in myotube formation *in vitro, J. Biol. Chem.,* 256, 8069, 1981.

120. **Puri, E. C., Chiquet, M., and Turner, D. C.,** Fibronectin-independent myoblast fusion in suspension cultures, *Biochem. Biophys. Res. Commun.,* 90, 883, 1979.

121. **Puri, E. C., Carvatti, M., Perriard, J. C., Turner, D. C., and Eppenberger, H. M.,** Anchorage-independent muscle cell differentiation, *Proc. Natl. Acad. Sci. U.S.A.,* 77, 5297, 1980.

122. **Schubert, D. and LaCorbiere, M.,** Role of a 16S glycoprotein complex in cellular adhesion, *Proc. Natl. Acad. Sci. U.S.A.,* 77, 4137, 1980.
123. **Schubert, D. and LaCorbiere, M.,** Properties of extracellular adhesion mediating particles in a myoblast clone and its adhesion-deficient variant, *J. Cell Biol.,* 94, 108, 1982.
124. **Prives, J. and Shinitzky, M.,** Increased membrane fluidity precedes fusion of muscle cells, *Nature (London),* 268, 761, 1977.
125. **Fulton, A. B., Rives, J., Farmer, S. R., and Penman, S.,** Developmental reorganization of the skeletal framework and its surface lamina in fusing muscle cells, *J. Cell Biol.,* 91, 103, 1981.
126. **Kalderon, N. and Gilula, N. B.,** Membrane events involved in myoblast fusion, *J. Cell Biol.,* 81, 411, 1979.
127. **Couch, C. B. and Strittmatter, W. J.,** Rat myoblast fusion requires metalloendoprotease activity, *Cell,* 32, 257, 1983.
128. **Chiquet, M. and Fambrough, D. M.,** Chick myotendinous antigen. I. A monoclonal antibody as a marker for tendon and muscle morphogenesis, *J. Cell Biol.,* 98, 1926, 1984.
129. **Chiquet, M. and Fambrough, D. M.,** Chick myotendinous antigen. II. A novel extracellular glycoprotein complex consisting of large disulfide-linked subunits, *J. Cell Biol.,* 98, 1937, 1984.
130. **Venkatasubramanian, K. and Solursh, M.,** Chemotactic behaviour of myoblasts, *Dev. Biol.,* 104, 428, 1984.
131. **Shellswell, G. B., Bailey, A. J., Duance, V. C., and Restall, D. J.,** Has collagen a role in muscle pattern formation in the developing chick wing?, *J. Embryol. Exp. Morphol.,* 60, 245, 1980.
132. **Hinchliffe, J. R. and Johnson, D. R.,** *The Development of the Vertebrate Limb,* Oxford University Press, England, 1980.
133. **Mauger, A., Kiney, M., Hedayat, I., and Goetnick, P. F.,** Tissue interactions in the organization and maintenance of the muscle pattern in the chick limb, *J. Embryol. Exp. Morphol.,* 76, 199, 1983.
134. **Lee, J. C.,** Electron microscope observations on myogenic free cells of denervated skeletal muscle, *Exp. Neurol.,* 12, 123, 1965.
135. **Reznik, M.,** Origin of the myogenic cell in the adult striated muscle of mammals, *Differentiation,* 7, 65, 1976.
136. **Mauro, A.,** Satellite cells of skeletal muscle fibres, *J. Biophys. Biochem. Cytol.,* 9, 493, 1961.
137. **Vanderburgh, H. H., Sheff, M. F., and Zacks, S. I.,** Soluble age-related factors from skeletal muscle which influence muscle development, *Exp. Cell Res.,* 153, 389, 1984.
138. **Bischoff, R.,** Regeneration of single skeletal muscle fibres *in vitro, Anat. Rec.,* 182, 215, 1975.
139. **Bischoff, R.,** Tissue culture studies on the origin of myogenic cells during muscle regeneration in the rat, in *Muscle Regeneration,* Mauro, A., Ed., Raven Press, New York, 1979, 13.
140. **Snow, M. H.,** Origin of regenerating myoblasts in mammalian skeletal muscle, in *Muscle Regeneration,* Mauro, A., Ed., Raven Press, New York, 1979, 91.
141. **Schiaffino, S., Pierobon Bormioli, S, and Aloisi, M.,** Fibre branching and formation of new fibres during compensatory muscle hypertrophy, in *Muscle Regeneration,* Mauro, A., Ed., Raven Press, New York, 1979, 177.
142. **Bruni, C.,** Mitotic activity of muscle satellite cells during the early stages of rhabdomyosarcomas induction with nickel sub-sulfide, in *Muscle Regeneration,* Mauro, A., Ed., Raven Press, New York, 1979, 265.
143. **Bischoff, R.,** Myoblast fusion, in *Cell Surface Rev.,* 5, 153, 1978.

Chapter 6

MORPHOGENESIS IN SPONGES

I. INTRODUCTION

The sponges (Porifera) are an aberrant line of evolution. Diversification within the phylum Porifera is not very spectacular. Sponges and their developmental phases do not appear to offer any phylogenetic "links" to connect ancestral unicellular organisms with the other phyla of multicellular animals. Development of sponges is a relatively simple process involving some peculiar larval forms. In its details, the developmental pattern seems to be basically different from that found in the other phyla of multicellular animals. For this reason, the Porifera are regarded even as a separate subkingdom, "Parazoa", of animals so as to indicate the absence of even remote phylogenetic relationships with the other multicellular forms.

However, the sponges have been a favorite experimental material with developmental biologists. The initial interest in sponges was simply for historical reasons. Wilson[1] observed that sponges can be disaggregated into single cells by just passing them through a fine mesh such as bolting silk. The disaggregated cells, when made to come together, eventually organize themselves into a mass constituting a functional sponge. Such sponges reassembled from disaggregated cells have typical histological structure characteristic of the species. Besides, when cells obtained from two different species of sponges (characterized by different colors) are mixed together, they segregate and form two separate sponges corresponding to their distinct origins. The classical work of Wilson[1] opened a new avenue of approach to understanding the mechanism of cell adhesion and aggregate formation. Many investigators have studied the formation of tissues from the disaggregated cells of the embryos of higher vertebrates. Such studies are more likely to reveal the mechanisms of development of the embryos of higher animals. These studies will be the subject of the next chapter. However, interest in the reconstitution of sponges has continued, and some very interesting facts regarding cell adhesion have been discovered during recent years.

II. STRUCTURAL ORGANIZATION OF SPONGES

Morphologically, most sponges are encrustations of irregular shape growing on submerged solid substrata, though there are some species that exhibit striking radial symmetry of "individual" shape. Histological complexity of the sponges varies considerably. For a detailed account, texts on sponges may be consulted.[2-4] For the present purpose, a brief description shall suffice. Typically, the sponge body consists of a number of interconnected cavities and canals organized into a structural unit so as to permit a flow of water through numerous pores on the surface into deeper cavities and eventually passing out. This constitutes the "canal system" of the sponges. A constant flow of water through the canal system carries with it tiny food organisms, which are trapped by specialized cells called choanocytes or collar flagellate cells. Water is employed to serve the same function as the internal body fluids of higher animals.

Considerable evolutionary divergence in the canal system has occurred. The location of the collar flagellate cells and the arrangement of canals and their branches show a wide degree of variation. Since these features are characteristic of the different subgroups of sponges, they are used as diagnostic features in taxonomy. The sponge body consists of some characteristic cell types and extracellular materials. The outer surface and the internal cavities (except those lined by the collar flagellate cells) are covered by an epithelial lining,

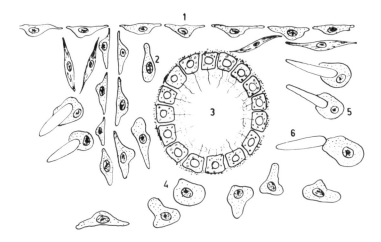

FIGURE 1.   Generalized diagram to show the different types of cells in sponges. 1, Pinacocyte; 2, grey cell; 3, canal surrounded by collar flagellate cells (choanocytes); 4, archeocyte; 5, scleroblast; and 6, spicule.

usually called the pinacoderm. The cells of the pinacoderm, the pinacocytes, are flat and polygonal with centrally bulged nucleated portions. In the pinacoderm are scattered pore cells through which the water flows in or out. The pinacoderm is an "epithelial layer" only in a general sense. It does not possess a basement membrane. Between the layers of the pinacoderm is a loose "connective tissue", usually called mesohyl. It contains a variety of cells and intercellular material. The cells of the mesohyl include the following subtypes: (1) archeocytes, which are described as totipotent amoeboid cells capable of differentiating into other cell types; (2) scleroblasts, which secrete the skeletal elements, viz., spicules or spongin fibers, and (3) other cell types such as collencytes, spongoblasts, gray cells, etc.[2]

Finally, the choanocytes (collar flagellate cells), which are characteristic of the sponges, are restricted to some parts of the internal cavities. Figure 1 illustrates the generalized histological structure of sponges. The intercellular material consists of fibrous (collagenous) and amorphous proteins. In the intercellular material are also found calcareous or silicious spicules of a spectacular variety of forms. In some sponges, the spicules are absent. The pinacocytes have a "surface coat" of glycoproteins and acid mucopolysaccharides.[5]

Experimental studies on sponges have been chiefly directed towards understanding the mechanism of mutual adhesion and morphogenetic rearrangement of cells. As mentioned above, disaggregated cells are capable of restoring typical histological organization, presumably through a mechanism involving cell adhesion and locomotion in addition to recognition of specific cell types. An important question in this context is whether all the types of sponge cells are necessary in order to bring about normal histological reorganization. It seems that so far this question has not received a satisfactory answer. We shall discuss here some aspects of adhesion and cell type recognition in sponges. An interesting question, which has received considerable attention, is regarding the individuality of sponges. This will also be alluded to.

III. CELL ADHESION

When mechanically disaggregated cells are allowed to settle down to the bottom of a container, they come together by their locomotory activity and adhere to each other. In stirred suspensions, the cells come together through collision and establish adhesive bonds. Experimentally, the progress of reaggregation can be monitored by the various methods described in Chapter 2 (Volume I).

A. The Aggregation Factor

Several workers have reported that sponge cells obtained by mechanical disaggregation reaggregate faster in the disaggregation supernatant than in plain sea water. This observation clearly indicates that the process of disaggregation results in the release of some cementing substance into the medium. The release is aided by low calcium media. When chelators (EDTA, etc.) are used for disaggregation, the cells lose their adhesiveness. Reaggregation of mechanically disaggregated cells is very slow in calcium-free media. From these observations, it could be concluded that the adhesive mechanism is Ca^{2+}-dependent. Further investigations have led to the identification of an "aggregation factor" released during disaggregation of the cells.

Generally it has been observed that the aggregation factor is species specific; i.e., it does not promote the aggregation of the cells of a different species. The factor obtained from the marine sponge, *Microciona prolifera,* enhances the reaggregation of homotypic cells in stirred suspensions. Cells of another marine sponge, *Haliclona occulata* do not aggregate in the presence of the *M. prolifera* factor. When a mixture of the cells of both the species was stirred in the presence of the *M. prolifera* factor, only the homotypic cells aggregated.[6] These observations demonstrate clearly that the action of the aggregation factor is species specific. McClay[7] investigated the species-specific effects of the aggregation factors from five different species belonging to four genera of sponges. The reaggregation rate of cells in the presence of the factor was compared with that in filtered sea water (controls). The factor had the effect of increasing the rate of reaggregation of homospecific cells. The factor had no effect on the reaggregation of heterospecific cells in most cases. In some cases, there was an inhibitory effect, whereas in other cases, the effect was somewhat ambiguous and not pronounced (Figure 2). These observations show that the adhesion mechanism in the different species is similar in that it consists of an aggregation factor that is released during cell dissociation. However, qualitatively, the factors are distinct.

Even formalinized cells show species-specific adhesion, indicating that the specificity lies in the cell surface molecules and is not related to the metabolic activities of the cells.[8] In some cases, the factor from one species can promote the aggregation of some heterospecific cells, albeit a little less efficiently.[9] The aggregation factor obtained from *Microciona prolifera* is a high molecular weight substance ($\approx 21 \times 10^6$ daltons) and has a "sunburst" configuration with a central "ring" and about 15 radiating "arms" as revealed by electron microscopy. Chemically the factor is a proteoglycan consisting of about equal proportion of protein and carbohydrate.[10] It has been shown that the factor is inactivated by β-glucuronidase. Glucuronic acid acts as a hapten-like inhibitor of cell aggregation in the presence of the factor. From these facts, Burger and associates concluded that the aggregation factor is recognized by a carbohydrate-specific mechanism on the cell surface. For details and references to earlier literature, see Burger et al.[8] and Turner.[11] In the fresh water sponge, *Spongilla carteri,* the carbohydrates involved in recognition are mannose and glucuronic acid.[12] It has been observed that chelators such as EDTA inactivate the aggregation factor irreversibly. This inactivation is observed as the loss of the "sunburst" configuration of the factor. Increasing concentrations of the chelator result in a corresponding decrease in the molecular weight of the factor, accompanied by the loss of its biological activity. Finally, the factor dissociates irreversibly into subunits of about 2×10^5 daltons.[8,13]

B. The Base Plate

It was mentioned above that the sponge aggregation factors are species specific. In other words, the factor from a given species will promote the aggregation of the cells from the same species but not of others. It could be inferred from this that the disaggregated cells have some surface features that characterize them. The factor can then be considered as a multivalent cementing molecule, which can recognize some molecules on the surface of the

Factor / Cells	Haliclona variabilis	Haliclona viridis	Tadania ignis	Homaxinella rudis	Dysidea crawshayi
Haliclona variabilis	+	−	⊕	⊖	○
Haliclona viridis	⊖	+	○	−	○
Tadania ignis	○	⊖	+	−	○
Homaxinella rudis	○	−	○	+	○
Dysidea crawshayi	○	⊖	○	⊖	+

FIGURE 2.   Aggregation behavior of sponge cells of different species in the presence of the aggregation factor from different species. (From McClay, D. R., *J. Exp. Zool.*, 188, 89, 1974. With permission.)

homospecific cells. These recognition molecules on the cell surface have been called the base plate. A variety of conventional methods of releasing cell surface proteins have failed to liberate the base plate from sponge cells. However, a controlled exposure of the cells of *Microciona* to hypotonic shock releases the base plate. Cells exposed to the hypotonic shock are viable after the release of the base plate,[14] suggesting that the treatment does not damage the cells. The lipid bilayers of such cells are intact, and from this it can be inferred that the base plate is neither a cytoplasmic component nor an integral membrane molecule. It is nondialyzable, pronase sensitive, stable to lipid extraction, and unaffected by EDTA. It tolerates a wide range of pH, from 3 to 12. The molecular weight of the base plate of *Microciona prolifera* has been estimated to be between 45,000 and 60,000 daltons. It does not seem to be similar to the aggregation factor or its subunits.

C. The Molecular Mechanisms of Sponge Cell Adhesion

When cells exposed to the hypotonic shock are stirred with the aggregation factor, they fail to aggregate. The substance released during the hypotonic shock is necessary for the cells to be aggregation-competent. In other words, cells with intact base plate can aggregate in the presence of the aggregation factor. From these observations, it is clear that the adhesion of sponge cells involves at least two components, viz., the base plate and the aggregation factor. A number of elegant experiments by Burger and associates have established this for *Microciona* and some other sponges. The process of sponge cell adhesion could be demonstrated in a simple model system. The aggregation factor was bound covalently on agarose beads, which could aggregate in the presence of Ca^{2+}. If the base plate was similarly bound on agarose beads, they could not aggregate, but could readily do so when the aggregation factor and Ca^{2+} were also added. Base plate-coated beads in the presence of glucuronic acid could not aggregate when the factor and Ca^{2+} were added. Mechanically disaggregated cells (with their base plate and aggregation factor intact on their surface) could adhere to the base plate-coated beads (see Figure 3). These elegant experiments provide an acellular model of sponge cell aggregation.

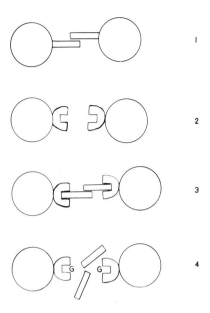

FIGURE 3. Aggregation of agarose beads conjugated with base plate or aggregation factor. The circles represent the agarose beads with the aggregation factor (rectangular bar) or base plate (depicted by grooved semicircles). Calcium ions (not shown) are required for the aggregation. 1, Beads with the aggregation factor adhere to each other; 2, beads with the base plate conjugated do not adhere even in the presence of calcium; 3, beads conjugated with the base plate and added aggregation factor adhere to each other; and 4, beads with conjugated base plate and added aggregation factor fail to adhere in the presence of glucuronic acid (G). (Based on the experiments of Burger, M. M., Burkart, W., Weinbaum, G., and Jumblatt, J., *Symp. Soc. Exp. Biol.*, 32, 1, 1978.)

The precise pattern of distribution of the base plate on the cell surface is not known. However, Kartha and Mookerjee[15] believe that there must be a definite pattern of distribution of such molecules on the cell surface. They find that agents that depolymerize microtubules cause the formation of larger aggregates. They suggest that this could be due to an altered distribution of the cell surface molecules involved in adhesion. This interesting suggestion, however, needs the strength of further experimental evidence. Since the base plate is, in all probability, a peripheral protein, how its distribution could be altered by interfering with the cytoskeletal elements is not easily understood. It is tempting to suggest that there are integral membrane proteins that serve to anchor the cytoskeletal elements as well as the base plate. However, nothing would be gained by discussing such a mechanism unless this possibility is tested experimentally.

The chemical nature of the base plate and aggregation factor differs in the various species studied. Unlike the *Microciona prolifera* factor, which has equal quantities of protein and carbohydrate, that of *Geodia cydonium* is predominantly protein. Nevertheless, the general mechanism of cell adhesion seems to be similar in consisting of two components, viz., a cementing substance (the factor) and a cell surface receptor (the base plate). Figure 4 depicts the manner in which the molecular mechanism holds the cells together.

D. Dual Mechanisms of Cell Adhesion

Working on *Geodia cydonium*, Müller[16,17] has described an additional adhesive mechanism. In this species, there is a mechanism that consists of a soluble adhesion factor and a membrane-associated complementary molecule called the aggregation receptor. The soluble factor, which acts as a linking molecule, is released into the disaggregation medium when the sponge tissue cells are separated in the presence of EDTA. The native factor is a macromolecular complex consisting of a core structure and three functional subunits: (1)

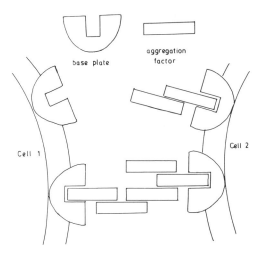

FIGURE 4.   Diagram showing the mechanism of sponge cell adhesion
mediated by the base plate and aggregation factor. Free aggregation factor
molecules can bind with each other and also with the base plate which is
a plasma membrane associated component.

glucuronosyl transferase, (2) galactosyl transferase, and (3) a subunit, which carries the
binding site to the cell surface membrane. The subunit with the cell surface affinity has C-
terminal lysine or arginine residues that have a functional role in binding. It has been inferred
that the adhesion factor corresponds to the aggregation factor of *Microciona*. The aggregation
receptor of *G. cydonium* is comparable with the base plate of *Microciona*. Müller[17] has
designated this as the secondary aggregation mechanism. The additional adhesive mechanism
described by Müller is designated as the primary adhesive mechanism. It works in a wide
range of temperature (2 to 30°C), pH (6.5 to 9.0), and ionic strength (0.05 M to 0.7 M
NaCl). The primary mechanism alone is not species specific. It consists of three subunits,
with the following molecular weights: 16,000, 15,000, and 13,500. Cell adhesion mediated
through the primary mechanism is calcium-dependent. The primary adhesion factor is firmly
associated with the plasma membrane, and no intermediate linking molecules are involved
in its function. Adhesion of cells utilizing the two mechanisms is shown in Figure 5.

In the hexactinellid sponge *Aphrocallistes vastus,* a different mechanism has been described.[18] This is distinguished from the adhesion factor of *G. cydonium* or that of *M.
prolifera* by an important criterion. The factor obtained from *A. vastus* was found to cause
nonspecies-specific cell aggregation (see Table 1). Particularly striking is the fact that even
mammalian cells from different established cell lines respond to the factor. Its biological
activity is not dependent on ionic strength, pH, and temperature within a broad range. The
cell binding sites carry no C-terminal lysine or arginine residues. Müller et al.[18] suggest that
two or more molecules of the *Aphrocallistes* adhesion factor (linked by Ca^{2+}) act as a bridge
between two cells.

E. The Role of Calcium Ions

The role of Ca^{2+} in the adhesion of cells is variable with respect to the different types
of aggregation mechanisms. The "primary" adhesion mechanism is Ca^{2+}-dependent. In
case of the "secondary" mechanisms, which consist of a linking molecular aggregate (the
factor) and a cell surface-bound receptor (the base plate), three categories can be distinguished
on the basis of the information available at present on different sponges.

1.   In *M. prolifera,* Ca^{2+} are required for the polymerization of the aggregation factor

primary aggregation mechanism

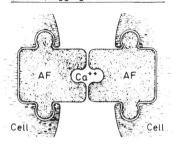

secondary aggregation mechanism

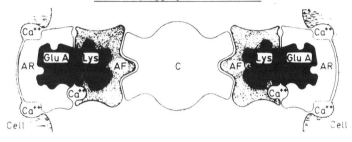

FIGURE 5. Diagrammatic representation of the two molecular mechanisms proposed to explain primary and secondary aggregation mechanism in the sponge *Geodia cydonium*. AF, Aggregation factor; AR, aggregation receptor; C, core structure; Lys, lysine; and GluA, glucuronic acid. (From Müller, W. E. G., *Prog. Clin. Biol. Res.*, 151, 359, 1984. With permission.)

subunits, constituting the multivalent factor ("sunburst" configuration). No Ca^{2+} are required for the interaction between the factor and the base plate. The interaction between the cell surface and the base plate also seems to be Ca^{2+}-independent.

2.   The aggregation factor of *G. cydonium* can bind the surface receptor (equivalent of the base plate) only in the presence of Ca^{2+}. The association of the receptor with cell surface is also Ca^{2+}-dependent.

3.   In *A. vastus*, two (or more) subunits of the soluble factor are linked by Ca^{2+}. (The nature of the interaction between the cell surface and the factor is not known at present.)

Do the different types of cells in the sponge body use the same mechanism (or mechanisms) for their adhesion or do they have cell type specific and distinct mechanisms for mutual adhesion? The importance of this question is obvious. Specific cell associations could be controlled by distinct mechanisms on the different types of cells. It is conceivable that during reaggregation the initial adhesion of cells, which seems to be essentially a physicochemical process, manifested even by killed cells, is followed by another, qualitatively different mechanism assuming a role in reorganizing the cells spatially. In other words, the mechanisms of adhesion and cell sorting could be distinct. In case all the cell types possess identical mechanisms, it will be difficult to account for sorting out resulting in the restoration of histological structure in the cell aggregates. It is possible to approach this problem if one can separate the different types of cells on a preparative scale. A few attempts have been made to achieve this.[8,19] The crude cell suspension prepared by mechanical disaggregation usually consists of archeocytes, choanocytes, grey cells, and rhabdiform cells. John et al.[20] obtained a biphasic separation of mechanically disaggregated cells by Ficoll density gradient centrifugation. Two species of marine sponges (*Ophlitaspongia seriata* and *Halichondria panicea*) were studied. Both gave archeocyte and mucoid cell-enriched fractions. These cell

Table 1
AGGREGATION POTENTIAL OF *APHROCALLISTES* AGGREGATION FACTOR TO HOMOLOGOUS AND HETEROLOGOUS CELLS

	Size of aggregates (μm)		
Cells	**ASW**	**ASW plus AF**	**CMF-SW plus AF**
Sponges:			
Aphrocallistes vastus	60 ± 10	2380 ± 420	95 ± 20
Clathrina coriacea	70 ± 10	1930 ± 300	80 ± 10
Geodia cydonium	70 ± 10	2550 ± 470	105 ± 30
Tethya lyncurium	70 ± 10	2770 ± 510	120 ± 30
Axinella polypoides	125 ± 20	1510 ± 280	155 ± 45
Mycale massa	55 ± 10	1230 ± 200	90 ± 30
Adocia grossa	140 ± 25	2250 ± 360	130 ± 40
Spongia officinalis	60 ± 10	1930 ± 310	75 ± 25
Mammalian cell lines:			
L5178y	20 ± 10	1450 ± 220	60 ± 15
X2	20 ± 10	1650 ± 270	60 ± 15
P815	20 ± 10	1310 ± 210	65 ± 20

Note: ASW, calcium- and magnesium-containing artificial sea water; CMF-SW, calcium- and magnesium-free artificial sea water; AF, aggregation factor. The AF was added at a concentration of 20 units/3 mℓ assay to 75 ± 25 × 10^6 cells and incubated under standard conditions in the presence of ASW or CMF-SW for 30 min. In the control assay, no AF was added. Each value represents the mean ± SD of five parallel assays.

From Müller, W. E. G., Conrad, J., Zahn, R. K., Steffen, R., Uhlenbruck, G., and Müller, I., *Differentiation*, 26, 30, 1984. With permission.

types differed in their aggregation characteristics. The relative proportion of the various cells was, however, variable according to season and other factors. Burger et al.[8] obtained enriched fractions of choanocytes by the method of velocity sedimentation using a Ficoll density gradient. The cells separated in this manner were allowed to adhere to sepharose beads with the aggregation factor bound covalently. This assay did not reveal preferential adhesion of any cell type to the sepharose beads. It seems that different methods of cell separation (e.g., isoelectric focusing, phase partition, and affinity chromatography) will have to be used in order to achieve a better separation of the cell types and to answer the question posed above.

F. Development of Junctional Complexes

The cellular architecture of sponges indicates a simple state of tissue organization. Cells are surrounded by an extracellular matrix and generally plasma membranes of adjacent cells are not in close apposition. Electron microscopic studies reveal that cell junctions show little specialization. Between the cells, which are closely apposed, there is a gap of 100 to 300 Å.[21,22] Larger gaps, measuring up to 1000 Å are also observed. The gaps are filled by an amorphous material, which is continuous with the mesohyl ground substance and the mucopolysaccharide cell coat. Kartha and Mookerjee[23,24] studied the ultrastructure of the fresh water sponge *Spongilla carteri* and observed loosely arranged cells embedded in the ground substance. Archeocytes were found to possess long, thin pseudopodia. Often the pseudopodial extensions exhibited dovetailing with other similar extensions. Membranous vesicles interpose between amoebocytes. Such vesicles are found located in the matrix, even away

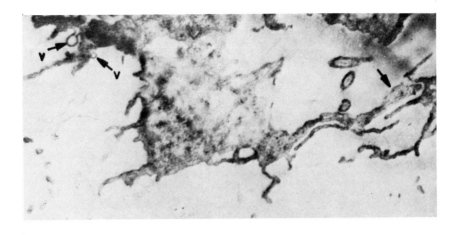

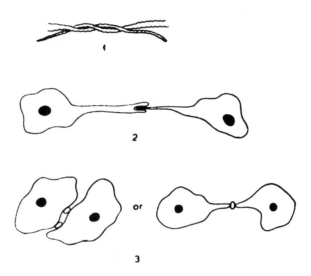

FIGURE 6. Above: Electron micrograph showing pseudopodial extensions of archeocytes of *Spongilla carteri* making dovetailing arrangements with each other (arrow). Vesicles (V) in the process of pinching off the surface may be seen. Below: Diagram representing different categories of cell contacts between sponge cells. 1, Contact established by intertwining filopodial extensions; 2, dovetailing contact; and 3, contact mediated by vesicles either between cells lying closely apposed or between pseudopodial extensions, the vesicles being interposed. (From Kartha, S. and Mookerjee, S., *Indian J. Exp. Biol.*, 16, 865, 1978. With permission.)

from the cells. The origin of these vesicles is probably the blebs formed at the surface of cells and pinched off to form free structures. The membranous vesicles are found between pinacocytes and amoebocytes also. Nothing is known about the function of such vesicles in the sponge body.

The pinacocytes show another interesting form of intercellular association. Thin, flat extensions from two adjoining cells are often found intertwining. The various types of cell associations observed by Kartha and Mookerjee[23] are depicted in Figure 6. When cell aggregates were examined ultrastructurally, it was found that the surfaces of the constituent cells undergo characteristic changes. Pinacocytes show extensive elongation at the periphery

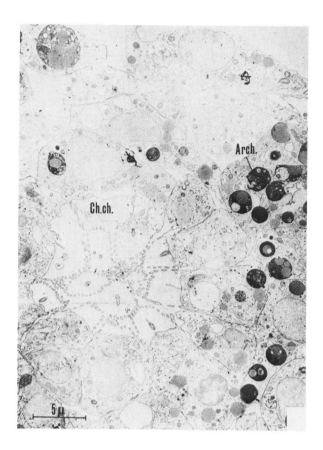

FIGURE 7.   An aggregate of the cells of *Ephydatia fluviatilis* cultured
for 24 hr showing the reorganization of a choanocyte chamber (Ch. ch.)
surrounded by other cells including archeocytes (Arch.). (From Van de
Vyver, G., *Curr. Top. Dev. Biol.*, 10, 123, 1975. With permission.)

and the projections from neighboring pinacocytes overlapping over a considerable length.
Archeocytes also establish dovetailing contacts as well as juxtapositions with intervening
vesicles. Kartha and Mookerjee[23,24] observed these changes in the aggregates of cells after
about 24 hr from reaggregation. Their observations indicate that the plasma membrane of
the sponge cells undergoes ''morphogenetic'' modifications after the formation of aggregates.
However, the possible relationship of these structural changes with the molecular mechanisms
of cell adhesion and rearrangement remains obscure. Whether the modifications are in any
way related to restoring the spatial order of cells in the aggregates is also a moot point.

The cells included in the aggregates are arranged randomly. Gradually they organize
themselves into definite tissue structures[25] (see Figure 7). This is an expression of tissue-
type specificity, which may be different from the species-specific adhesion specificity. Such
a conclusion may be arrived at from a study of the pattern of reorganization in the aggregates
of *Spongilla carteri* cells also.[26]

IV. SPECIES-SPECIFIC RECOGNITION AND INDIVIDUALITY IN SPONGES

The biological significance of the existence of species specificity in sponges is readily
understandable. Sponges are organisms (colonies?), which grow on submerged rocks and
such other solid supports. In the natural environment, several species of sponges may be

growing in the same locality, and if the mass of one grows into another, a mixed heterospecific sponge would result. Of course, this does not happen. However, can one mass of a given species grow into another of the same species resulting in a larger individual or a complex colony? If this did happen, any species of sponge could be considered as a gigantic "individual" or colony fragmented and distributed in different localities! Of course, this is unthinkable. *A priori,* one would expect that there is individuality in sponges. However, its existence could be established only by definite experimental investigations.

Moscona[27] and McClay[7] reported experimental studies that gave a clear indication that individuality exists in sponges. When allografts (grafts of different individuals of the same species) are made, they do not heal, whereas homografts (grafts from the same individual) are found to heal readily. Even in nature where two masses of the same species of sponges are growing abutting each other, there is always an intervening zone of necrotic cells, and the two masses never have a contiguity of live tissues. It could be suggested that the absence of confluence between two growing individuals could be due to the lack of adhesion between the exposed surfaces of pinacocytes. Alternatively it could be owing to the pinacocytes of one individual recognizing the same type of cells of another individual as alien. However the rejection of allografts indicates clearly that individuality exists in sponges, and it is expressed in all tissues of the sponge. Allograft rejection has been studied, using conventional histological methods as well as transmission and scanning electron microscopy. Van de Vyver and Barbieux[28] studied the process in two species of the marine sponges of the genus *Polymastia.* In *P. robusta,* an allograft is rejected by the formation of a boundary between the intervening tissues. A few days after making a graft, archeocytes and collencytes migrate toward the zone of contact where they align on either side of the boundary. A collagenous fibrous material is elaborated at the intervening zone. The two tissues may remain together with the intervening zone of the "scar". In *P. mamilaris,* allograft rejection is more complete. The graft never fuses with the host. There is no alignment of the cells that migrate toward the boundary. There is also no deposition of collagenous material. Thus the two species differ in the manner in which allografts are rejected.[28] The cellular aspects of allogenic recognition have also been studied in three different species of the genus *Axinella.*[29,30] In these forms, archeocytes migrate invasively into the zone of alien tissue. Extensive phagocytosis and cell lysis are observed. The cytotoxic effects during the host-graft interaction seem to be working both ways, i.e., against the graft as well as the host.

An interesting observation has been made by Van de Vyver[25] who studied the fresh water sponge *Ephydatia fluviatilis.* If two or more gemmules of the sponge germinate very close to each other, they can merge into a single mass. However, sometimes this does not happen, and the masses developed from the gemmules grow as separate individuals. A detailed study of this species of fresh water sponge has revealed the existence of several distinct strains. Gemmules of the same strain can fuse, whereas those of different strains do not.[25]

Van de Vyver[25] also observed that the aggregation factor (a thermolabile substance of $\approx 50,000$ mol wt) from the distinct strains differs. Thus the factor from a given strain could promote the aggregation of the cells of its own type but not of another strain. The strain specificity thus seems to reside in the adhesive mechanism. Allogenic incompatibility in the different strains of *Ephydatia fluviatilis* has been described recently by Buscema and Van de Vyver.[31] The interaction between the allograft and host is characterized by the formation of a barrier secreted at the interface. Ultrastructural studies revealed that the barrier consists of collagen and spongin fibers. Intense incorporation of tritiated proline in the barrier material further shows that it is composed of collagen. The involvement of several cell types in the barrier formation has been indicated.

V. CONCLUDING REMARKS

Sponges will no doubt continue to be the objects of further studies on two aspects that are of general interest to all developmental biologists: cell adhesion and recognition. It appears that sponges have diversified their adhesive mechanisms during evolution. As discussed above, there are two distinct types of adhesive mechanisms: (1) those involving an intermediate linking macromolecular complex and a cell surface associated base plate (secondary mechanism); and (2) those in which there is no intermediate linking molecule, the cell surface associated mechanisms linking the cells directly through their complementary adhesive domains (primary mechanism). According to Müller,[17] the calcareous sponges are equipped with only the primary mechanism, whereas the others are provided with an additional, viz., the secondary mechanism. The adhesive mechanisms differ in their Ca^{2+} requirements also. Since the primary mechanism is not species specific, how the calcareous sponges recognize each other species specifically is not clear.

Sponges seem to have evolved recognition mechanisms working at three different levels: (1) the mechanism which recognizes the cell types and is responsible for their rearrangement in an aggregate; (2) the mechanism which recognizes individuals and is expressed in allograft rejection, and (3) the mechanisms which recognize different species.

REFERENCES

1. **Wilson, H. V.,** On some phenomena of coalescence and regeneration in sponges, *J. Exp. Zool.,* 5, 245, 1907.
2. **Hyman, L. H.,** *The Invertebrata,* Vol. 1, McGraw-Hill, New York, 1940, chap. 6.
3. **Barnes, R. D.,** *Invertebrate Zoology,* 3rd ed., W. B. Saunders, Philadelphia, 1974, chap. 4.
4. **Bergquist, P. R.,** *The Sponges,* Hutchinson, London, 1978.
5. **Garonne, R., Thiney, Y., and Pavans de Cecalty, M.,** Electron microscopy of a mucopolysaccharide coat in sponges, *Experientia,* 27, 1324, 1971.
6. **Humphreys, T.,** Chemical and *in vitro* reconstruction of sponge cell adhesion. I. Isolation and functional demonstration of the components involved, *Dev. Biol.,* 8, 27, 1963.
7. **McClay, D. R.,** Cell aggregation: properties of cell surface factor from five species of sponges, *J. Exp. Zool.,* 188, 89, 1974.
8. **Burger, M. M., Burkart, W., Weinbaum, G., and Jumblatt, J.,** Cell-cell recognition: molecular aspects. Recognition and its relation to morphogenetic processes in general, *Symp. Soc. Exp. Biol.,* 32, 1, 1978.
9. **Turner, R. S., Jr. and Burger, M. M.,** A two component system for surface-guided reassociation of animal cells, *Nature (London),* 244, 509, 1973.
10. **Henkart, P. S., Humphreys, S., and Humphreys, T.,** Characterization of sponge aggregation factor: a unique proteoglycan complex, *Biochemistry,* 12, 3045, 1973.
11. **Turner, R. S., Jr.,** Sponge cell adhesion, in *Specificity of Embryological Interactions,* Garrod, D. R., Ed., Chapman & Hall, London, 1978, chap. 6.
12. **Kartha, S. and Mookerjee, S.,** Reaggregation of sponge cells: specificity and recovery of adhesion, *Wilhelm Roux Arch.,* 185, 155, 1979.
13. **Jumblatt, J. E., Schlup, V., and Burger, M. M.,** Cell-cell recognition: specific binding of *Microciona* sponge aggregation factor to homotypic cells and the role of calcium ions, *Biochemistry,* 19, 1038, 1980.
14. **Jumblatt, J. and Burger, M. M.,** Studies on the binding of *Microciona prolifera* aggregation factor to dissociated cells, *Biol. Bull.,* 153, 131, 1977.
15. **Kartha, S. and Mookerjee, S.,** Microtubule depolymerizing drugs enhance cell reaggregation in sponges, *Indian J. Exp. Biol.,* 17, 439, 1979.
16. **Müller, W. E. G.,** Cell membranes in sponges, *Int. Rev. Cytol.,* 77, 129, 1982.
17. **Müller, W. E. G.,** Two distinct, functionally independent adhesion mechanisms in marine sponges, *Prog. Clin. Biol. Res.,* 151, 359, 1984.
18. **Müller, W. E. G., Conrad, J., Zahn, R. K., Steffen, R., Uhlenbruck, G., and Müller, I.,** Cell adhesion molecule in the hexactinellid *Aphrocallistes vastus, Differentiation,* 26, 30, 1984.

19. **Leith, A. G. and Steinberg, M. S.,** Velocity sedimentation, separation and aggregative specificity of discrete cell types, *Biol. Bull.,* 143, 468, 1972.
20. **John, H. A., Campo, M. S., Mackenzie, A. M., and Kemp, R. B.,** Role of different sponge cell types in species specific cell aggregation, *Nature (London) New Biol.,* 230, 126, 1971.
21. **Bagby, R. M.,** The fine structure of pinacocytes in the marine sponge *Microciona prolifera* (Ellis and Solander), *Z. Zellforsch.,* 105, 579, 1970.
22. **Bagby, R. M.,** Formation and differentiation of the upper pinacoderm in reaggregation masses of the sponge *Microciona prolifera* (Ellis and Solander), *J. Exp. Zool.,* 108, 217, 1972.
23. **Kartha, S. and Mookerjee, S.,** Ultrastructure of cell contact in sponge cells, *Indian J. Exp. Biol.,* 16, 865, 1978.
24. **Kartha, S. and Mookerjee, S.,** Cell contact in aggregating sponge cells. An ultrastructural study, *Mikroscopie,* 35, 213, 1979.
25. **Van de Vyver, G.,** Phenomena of cellular recognition in sponges, *Curr. Top. Dev. Biol.,* 10, 123, 1975.
26. **Kartha, S. and Mookerjee, S.,** Pattern of tissue reconstruction from sponge cell aggregates, *Naturwissenschaften,* 65, 599, 1978.
27. **Moscona, A. A.,** Cell aggregation: properties of specific cell ligands and their role in the formation of multicellular systems, *Dev. Biol.,* 18, 250, 1968.
28. **Van de Vyver, G. and Barbieux, B.,** Cellular aspects of allograft rejection in marine sponges of the genus *Polymastia, J. Exp. Zool.,* 227, 1, 1983.
29. **Buscema, M. and Van de Vyver, G.,** Cellular aspects of alloimmune reactions in sponges of the genus *Axinella.* I. *Axinella polypoides, J. Exp. Zool.,* 229, 7, 1984.
30. **Buscema, M. and Van de Vyver, G.,** Cellular aspects of alloimmune reactions in sponges of the genus *Axinella.* II. *Axinella verrucosa* and *Axinella damicornis, J. Exp. Zool.,* 229, 19, 1984.
31. **Buscema, M. and Van de Vyver, G.,** Allogenic recognition in sponges: development, structure and nature of the non-merging front in *Ephydatia fluviatilis, J. Morphol.,* 181, 297, 1984.

Chapter 7

CELL SORTING: AN EXPERIMENTAL MODEL OF MORPHOGENESIS

I. SORTING OUT IN EMBRYONIC TISSUE CELL AGGREGATES

Embryonic development is characterized by extensive migration of cells and the establishment of new cell associations. This initiates interactions among cells eventually leading to their differentiation. Even early embryos are complex systems with many different types of cells engaged in migration, interaction, and differentiation. An experimental analysis of such a complex system is therefore difficult. An easier avenue of approach would be to study simpler systems.

As described in the earlier chapters, the cellular slime molds and sponges offer interesting experimental material to reveal the cellular mechanisms of morphogenetic changes. The study of these simple model systems has clearly shown that certain generalized properties of the cell surface are directly involved in the development of pattern. In particular, the molecules making up the plasma membrane confer some properties that determine such cellular interactions as mutual adhesion, specific recognition, and migration with reference to one another. Simple experimental models consisting of parts of vertebrate embryos have also been studied by developmental biologists. The pioneering work of Holtfreter in the fifties initiated extensive research in this area.

Holtfreter used early amphibian embryos in his experiments.[1] A neurula could be dissected to obtain different cell layers such as the neural plate, epidermis, endoderm, etc. as shown in Figure 1. These were taken in various combinations and grown for a few to several days during which the pieces healed together into compact masses and showed rearrangements. Experiments were also performed in which the tissue pieces were first disaggregated into single cells and then combined by mixing them together in a random arrangement. In some experiments, for ease of identification, embryos of two different species (*Ambystoma punctatum* and *Triturus torosus*), which differ in pigmentation, were used. The resulting rearrangements observed by Townes and Holtfreter[1] are summarized in Table 1. These studies indicated that individual cells retain their tissue type properties and engage themselves in morphogenetic movements, acquiring positions often resembling those in whole embryos. However, sometimes they also show mutually reversed positions. In other words, in some of the experimental combinations of tissue fragments or disaggregated cells, the morphogenetic rearrangement is quite unlike their arrangement in an embryo. This indicates that a tissue can determine the position of another tissue in relation to its own position. This in particular was an important outcome of the study. When neurula endoderm and mesoderm were combined, they reversed their natural position, i.e., instead of the mesoderm enclosing the endoderm, the former was enclosed by the latter. The experiments also indicated that early embryonic cells have mechanisms by which they can recognize other cells of like and unlike type. The mutual affinities among cells differ according to their origin. An additional point that emerged from the study was that single cells by their motile properties can reestablish histological organization which is lost during experimental disaggregation.

Several workers have used embryonic tissue cells chiefly from avian and mammalian embryos for investigating morphogenetic cell sorting. Selected tissues from chick, mouse, or other embryos are dissected out, and they are finely chopped into small pieces. Free cells are obtained by disaggregating the tissue pieces in calcium-magnesium-free media with or without proteolytic enzymes or chelators. The cells of different tissues are then mixed in suitable proportion and obtained as a pellet by centrifugation or by aggregation in a stirred suspension. Pieces of the pellet or the aggregates are then placed on a suitable support such

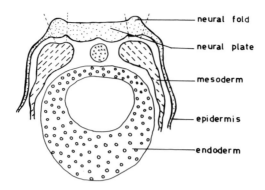

FIGURE 1.   Schematized transverse section of an amphibian neurula with different cell layers separated.

Table 1
MORPHOGENETIC REARRANGEMENT OF AMPHIBIAN NEURULA TISSUE FRAGMENTS AND CELLS IN VITRO

Tissues/cells combined[a]	Configuration after rearrangement[b]	Additional remarks[c]
A. Tissue fragments		
1. Medullary plate, epidermis	(Medullary plate) epidermis	(1)
2. Medullary plate, endoderm	(Medullary plate) endoderm	(2)
3. Medullary plate with neural fold, endoderm	[(Medullary plate) mesenchyme] endoderm	(3)
4. Epidermis, endoderm	Epidermis/endoderm	(4)
5. Prospective notochord, endoderm	(Notochord) endoderm	
6. Prospective somite, endoderm	(Somite mesoderm) endoderm	(5)
7. Prospective lateral plate, endoderm	(Mesoderm) endoderm	(5)
8. Whole mesoderm, endoderm	(Mesoderm) endoderm	(5); (6)
B. Disaggregated cells		
9. Medullary plate, epidermis	(Medullary plate) epidermis	(1)
10. Medullary plate, neural fold, epidermis	[(Medullary plate) mesenchyme] epidermis	
11. Medullary plate, endoderm	(Medullary plate) endoderm	(7)
12. Medullary plate, neural fold, endoderm	[(Medullary plate) mesenchyme] endoderm	(3)
13. Whole mesoderm, endoderm	(Mesoderm) endoderm	(5)
14. Epidermis, whole mesoderm	(Mesoderm) epidermis	
15. Lateral plate mesoderm, epidermis, endoderm	Medullary plate/epidermis/endoderm	(8)

[a]   Tissue fragments as shown in Figure 1.
[b]   Rearranged configuration is shown by brackets and parentheses. Enclosed cells within parentheses, when two tissue types are involved; where three-cell types are taken, the successive enclosing and enclosed phases are shown by brackets and parentheses.
[c]   Additional remarks: (1) No contact between the two. (2) The neural tissue after long durations of culture moved out and isolated itself from the mass of endodermal cells. (3) Rearrangement is not strictly concentric; the medullary fold gives rise to some mesenchyme and epidermis. (4) The two tissues remain as separate masses. (5) Note that the relative position of the two tissues is unlike in an embryo. (6) The "mesoderm" includes notochord also. (7) No self-isolation after prolonged culture. (8) No concentric rearrangement; the epidermis is never in contact with the endoderm without an intervening mesenchyme.

Based on Townes, P. L. and Holtfreter, J., *J. Exp. Zool.*, 128, 53, 1955.

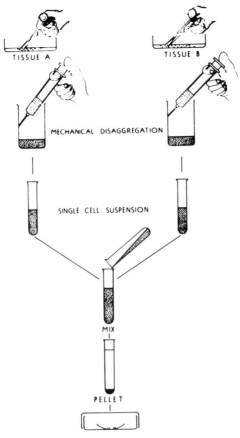

FIGURE 2. Experimental procedure followed in the study of cell sorting. (From Rao, K. V., Grover, A., and Beohar, P. C., *Prog. Clin. Biol. Res.*, 151, 345, 1984. With permission.)

as a millipore filter or the vitelline membrane of the hen's egg mounted around a glass ring (Figure 2). The cell aggregates are then incubated in a suitable nutritive medium for 3 to 4 days at the end of which they are examined histologically.

Typically, when two types of cells are mixed together and allowed to undergo rearrangement, it is found that one of the cell types forms a more or less compact core and the other surrounds it as a cortex (Figure 3). The rearrangement undergone by the cells is a morphogenetic reorganization in the sense that some histological features of the original tissue are developed again. When chick or rat embryonic kidney cells constitute one of the components of the cell sorting system, they form typical kidney tubules with their brush border ends surrounding a lumen. The tendency to form an epithelium is strongly expressed by the kidney cells even when they constitute the enclosing phase, in which case they form a continuous covering layer around the cell aggregate, with the brush border ends exposed. At certain places, the kidney epithelial cells invaginate to form vesicles or tubules of considerable length located deeper in the cell aggregate. Histological differentiation is exhibited by other types of embryonic cells also. When liver cells form the enclosed phase, they develop cavities reminiscent of bile ducts. Pigmented retina cells form small vesicles and rosettes that do not bear any resemblance to their normal tissue arrangement.

A few important features of the cell sorting model must be noted. The rearrangement of

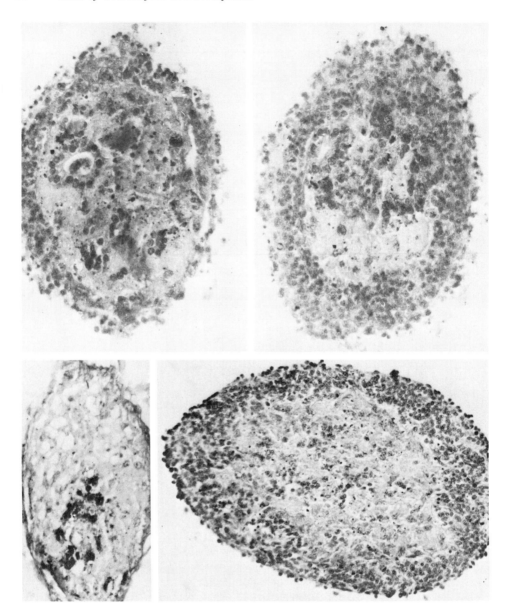

FIGURE 3. Sorting out in heterotypic cell aggregates. *Upper, left:* chick embryonic brain cells (enclosing phase) and rat embryonic kidney cells (enclosed phase). Some kidney cells have organized into a tubule enclosing some dead cells. *Upper, right:* chick embryonic neural retina cells (enclosing phase) and rat embryonic kidney cells (enclosed phase). *Lower, left:* chick embryonic leg muscle cells (enclosing phase) and chick embryonic pigmented retina cells (enclosed phase). *Lower, right:* chick embryonic neural retina cells (enclosing phase) and leg muscle cells (enclosed phase). (All magnification × 300.) (The lower right figure is from Grover, A. and Rao, K. V., *Cell Differentiation,* 13, 209, 1983. With permission.)

cells results in the segregation of the tissue types mixed. In other words, the tissue type affinity of the cells is strongly expressed and distinct homotypic cellular domains are formed. Typically the domains are continuous and concentric, giving rise to the enclosed and enclosing phases. In a few instances, however, the enclosed phase may be in the form of scattered discontinuous domains. Chick embryonic pigmented retina is generally found as scattered vesicles and rosettes in a continuous enclosing phase consisting of the other type of cells with which it is mixed.

Table 2
SORTING OUT IN CELL AGGREGATES[a]

No.

1.	Chick embryonic brain	(Chick embryonic pigmented retina)
2.	Chick embryonic neural retina	(Chick embryonic pigmented retina)
3.	Rat embryonic kidney	(Chick embryo fibroblasts)
4.	Chick embryo fibroblasts	(Chick embryonic pigmented retina)
5.	Chick embryonic heart	(Chick embryonic pigmented retina)
6.	Chick embryonic neural retina	(Chick embryonic kidney)
7.	Chick embryonic brain	(Chick embryo fibroblasts)
8.	Chick embryonic neural retina	(Chick embryonic heart)
9.	Rat embryonic brain	(Chick embryonic kidney)
10.	Rat embryonic brain	(Chick embryonic muscle)
11.	Rat embryonic brain	(Chick embryonic neural retina)
12.	Rat embryonic brain	(Chick embryonic liver)
13.	Rat embryonic brain	(Chick embryonic brain)
14.	Chick embryonic neural retina	(Chick embryo fibroblasts)
15.	Chick embryonic neural retina	(Chick embryonic liver)
16.	Rat embryonic brain	(Chick embryonic heart)
17.	Rat embryonic kidney	(Chick embryonic heart)
18.	Rat embryonic brain	(Chick embryo fibroblasts)
19.	Chick embryonic brain	(Chick embryonic kidney)
20.	Chick embryo fibroblasts	(Chick embryonic kidney)
21.	Chick embryonic neural retina	(Rat embryonic kidney)
22.	Chick embryonic brain	(Rat embryonic kidney)
23.	Chick embryonic heart	(Chick embryonic kidney)
24.	Chick embryonic liver	(Chick embryonic kidney)
25.	Rat embryonic kidney	(Chick embryonic liver)
26.	Chick embryonic liver	(Chick embryonic muscle)
27.	Chick embryonic liver	Chick embryonic heart[b]
28.	Chick embryonic muscle	(Chick embryonic kidney)
29.	Chick embryonic liver	(Chick embryonic pigmented retina)
30.	Rat embryonic kidney	(Chick embryonic muscle)
31.	Chick embryonic muscle	(Chick embryonic pigmented retina)
32.	Rat embryonic kidney	(Chick embryonic pigmented retina)
33.	Chick embryonic kidney	(Chick embryonic pigmented retina)
34.	Rat embryonic kidney	(Chick embryonic kidney)
35.	Chick embryonic brain	(Chick embryonic heart)
36.	Chick embryonic brain	(Chick embryonic muscle)
37.	Rat embryonic fibroblasts	(Chick embryonic muscle)
38.	Rat embryonic fibroblasts	(Chick embryonic brain)
39.	Chick embryonic liver	(Chick embryonic fibroblasts)
40.	Chick embryonic liver	Chick embryonic brain[b]

[a] Tissue type enclosed is in parentheses.
[b] Show reversal in tissue positioning.

A. Cell Type Hierarchy

The relative position assumed by a given cell type seems to depend on the other cells constituting the aggregate. A number of combinations in their rearranged configuration are given in Table 2. It can be seen readily from the table that a given cell type, which is the enclosed phase in some combinations, may constitute the enclosing phase in others. Thus

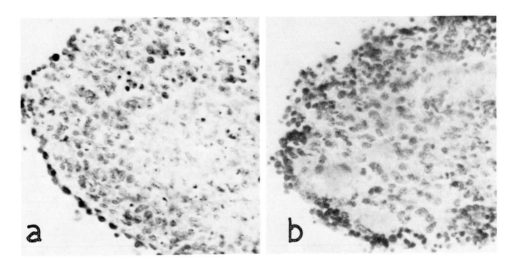

FIGURE 4.   Sorting out in a mixed aggregate of chick and rat embryonic brain cells. (a) The rat brain cells, characterized by larger nuclei, constitute the enclosing phase. (b) Heparin-treated rat brain cells assume the enclosed position. (Magnification ×400.) (From Grover, A. and Rao, K. V., *Cell Differentiation,* 13, 209, 1983. With permission.)

chick embryo leg muscle cells are enclosed when in combination with chick embryonic liver, brain, or rat embryonic kidney or rat embryo fibroblasts. On the other hand, the chick embryo leg muscle cells enclose chick embryonic pigmented retina or kidney cells. Also, the positioning does not seem to be related to the tissue type. Chick embryo fibroblasts, when combined with chick embryonic liver, muscle, brain, or rat embryonic kidney cells, constitute the enclosed phase, the other cell type, respectively, forming the enclosing phase. Rat embryo fibroblasts on the other hand constitute the enclosing phase when combined with these cells.

Sorting out patterns obtained from disaggregated embryonic cells derived from the same tissue but from different species have been studied by a number of workers. Moscona and collaborators, studying aggregates containing mouse and chick embryonic tissue cells, report that cells of the same type from the two sources do not sort out from each other. On the contrary, other workers have observed clear sorting out of tissues of the same type from two different species (Figure 4). These observations are not entirely in conformity with a generally held view that tissue type specificity is conserved even in distantly related species of warm blooded animals. Thus species specificity may be expressed in addition to the tissue specificity. Nag et al.[8] observed that chick and rat myocardial cells established intercellular contacts and junctions and formed synchronously beating hearts. Whether species specificity overrides tissue type specificity during cell sorting needs to be investigated further. At present, therefore, the question is open.

A seminal idea that explains cell positioning in sorting out came from Steinberg.[9] A striking tissue hierarchy in the behavior of sorting out in binary combinations was first discovered by Steinberg,[9] which was later confirmed by Grover et al.[7] and Grover and Rao.[10] The hierarchy is shown in Table 3. In the table, the tissues have been listed in such a manner that in a binary combination, any tissue would form the enclosing phase when combined with any other listed above it. The hierarchy is inferred from the observation of cell sorting in more than 40 binary combinations.[7,10]

B. The Pattern Emerging from Sorting Out

An important feature of the pattern obtained after sorting out should be noted. Most binary

Table 3
TISSUE HIERARCHY IN CELL SORTING

Hierarchical order	Source of the cells	Stage[a]
1	Chick embryonic pigmented retina	34
2	Chick embryonic kidney	33
3	Chick embryo fibroblasts grown in vitro (low passage)	
4	Chick embryonic heart	33
5	Chick embryo leg muscle	34
6	Chick embryo liver	34
7	Rat embryonic kidney	17 days
8	Chick embryonic brain	25
9	Chick embryonic neural retina	34
10	Rat embryonic brain	14 days

[a] Chick embryo stages are according to the scale of Hamburger and Hamilton.[42] In case of rat embryos, days of gestation are given (day of sperm detection taken as 0).

From Grover, A. and Rao, K. V., *Cell Differentiation*, 13, 209, 1983. With permission.

combinations of tissue cells made in the experiments resulted in bizarre assemblies of tissues. A configuration of kidney or liver cells enclosed within brain or neural retina cells is not found in any embryo. When chick neural and pigmented retina cells are combined, they sort out to form a continuous phase of the former cell type with scattered masses of the latter as an enclosed phase. This is quite unlike the spatial organization of the two tissues in the embryo. Finally, the pattern of one tissue enclosing another is not a normal organization of any organ except in rare instances like the adrenal gland. Therefore it is important to remember that the experimental model of sorting out cannot be assumed to be homologous to animal embryos. Nevertheless the experimental model has been studied with great interest with the hope that some of the cellular properties responsible for morphogenesis can be studied in the simpler model, albeit in a somewhat modified form. Besides, the modifications imposed experimentally may give clues to the altered morphogenetic behavior of the cells, thereby revealing the mechanisms underlying normal development.

II. THE MECHANISMS OF CELL SORTING

Several theories have been proposed to account for the sorting out behavior in heterotypic cell aggregates. Any discussion on this has to recognize two aspects of the phenomenon: (1) the cells segregate into distinct homotypic domains, and (2) they assume a definite position (i.e., enclosed or enclosing) in relation to the other cell type. It may be presumed that cells adhere due to the presence of complementary molecules at the surface. Recent work on this question (Volume I, Chapter 2) clearly implicates the glycoproteins of the plasma membrane as probable candidates for the role. Adhesion of like as well as unlike cells is possible, especially since the cell surface has more than one mechanism for establishing the contact. Glycoproteins from neural retina,[11,12] liver,[13] or brain[14] stimulate preferentially the adhesion of homotypic cells.

Clearly there is enough evidence now to show that specific adhesive mechanisms on the cell surface can account for the formation of homotypic cellular domains during sorting out. However, it is not equally clear if the same mechanisms can determine the relative position of the cell types after sorting out. It is possible that the avidity with which the cell surface ligands bind complementary molecules in the intercellular space or on the surface of other cells can vary quantitatively and thus result in different degrees of adhesiveness. There is at present no evidence to show that it is so; nonetheless, looking for such evidence may be rewarding. Adhesion of heterotypic cells at the zone of contact between the enclosed and enclosing phases is, in all probability, not specific. It is well-known that cells, which are never found together in vivo, may meet at the junction between the two phases. Sorting out of endoderm and ectodermal tissues of *Ambystoma* neurulae resulting in the isolation of the two tissues[1] is a case of total adhesive incompatibility. However, since a variety of unnatural cell combinations sort out into compact masses consisting of enclosed and enclosing phases, it seems difficult to account for cell sorting on the basis of specific adhesion alone.

A. Surface Free Energies and Cell Sorting

An attractive hypothesis to account for cell segregation as well as positioning of the cells in sorting out was proposed by Steinberg.[9,15] The tissue hierarchy noted earlier is obviously a reflection of some graded difference among the tissue cells. Perhaps it is some cell surface property that varies quantitatively. Sorting out in a wide variety of unnatural binary combinations indicates that the graded difference is of some general property of the cells rather than specific recognition or any other unique attribute of the cell types. Steinberg proposed that the hierarchy represents differences in adhesiveness of the different cell types. He also proposed that the more strongly adhesive cell type would constitute the enclosed phase.

Phillips and Steinberg[16] and Steinberg[17] have provided a detailed theoretical basis for the hypothesis. The factor that determines the reorganizational behavior of cells in sorting out is the specific interfacial free energy of the system (σ). It has been shown that cell aggregates exhibit some properties of liquid droplets. The equilibrium shape assumed by a liquid droplet in a gravitational field depends on its specific interfacial free energy (σ) with respect to the medium in which it is held. The higher the σ of the droplet, the rounder will be its profile at shape equilibrium. Let us consider the forces that determine the shape of two drops of immiscible fluids *a* and *b* held together in a medium *o* (Figure 5). The interfacial free energy between the droplet *a* and the medium is represented as σ_{ao}. Similarly, σ_{bo} and σ_{ab} represent the interfacial free energies at the interfaces *bo* and *ab,* respectively. They are shown as vectors by arrows in the diagram and are characterized by the relationship:

$$\sigma_{ao} + \sigma_{bo} > \sigma_{ab} > \sigma_{ao} - \sigma_{bo} \tag{1}$$

This set of σ values determines the equilibrium configuration shown in Figure 5, wherein droplet *b* envelops droplet *a* partially, and the two are held together in the medium *o*.

Further, it has been shown by Phillips and Steinberg[16] that cell aggregates assume an equilibrium shape under centrifugal force, and the shape assumed is independent of their original shape. Cell aggregates centrifuged at 4000 or 8000 *g* for 24 or 48 hr assume a characteristic shape as shown in Figure 6. The different shapes assumed by the pieces are a reflection of their σ. The shape of the component cells of such aggregates during centrifugation, however, remains similar to their native shape. This was shown in sections of the cell aggregates fixed during centrifugation. The three tissues illustrated in Figure 6 show a graded difference in their σ, and this corresponds to the tissue hierarchy observed by Steinberg.[9] This can be represented as

$$\sigma_{\text{limb bud}} > \sigma_{\text{heart ventricle}} > \sigma_{\text{liver}}$$

According to Steinberg,[17] the behavior of liquid droplets and cell aggregates is analogous. In case of the latter, however, it is easier to visualize the mutual adhesive strength of cells in terms of the reversible work of adhesion (W), which is defined as the reversible work required to separate a column of liquid of unit cross sectional area into two sections with an air phase to separate them. In this process, two surfaces of this area are formed and hence W is related to σ in the following manner:[17]

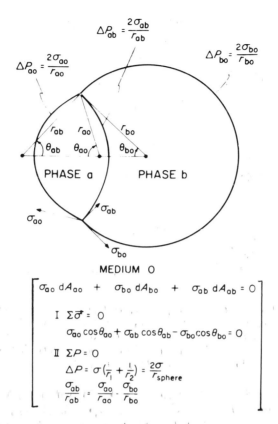

$$\begin{bmatrix} \sigma_{ao}\,dA_{ao} \;+\; \sigma_{bo}\,dA_{bo} \;+\; \sigma_{ab}\,dA_{ab} = 0 \\[6pt] \text{I } \Sigma\vec{\sigma} = 0 \\[4pt] \sigma_{ao}\cos\theta_{ao} + \sigma_{ab}\cos\theta_{ab} - \sigma_{bo}\cos\theta_{bo} = 0 \\[6pt] \text{II } \Sigma P = 0 \\[4pt] \Delta P = \sigma\left(\frac{1}{r_1} + \frac{1}{r_2}\right) = \frac{2\sigma}{r_{sphere}} \\[6pt] \frac{\sigma_{ab}}{r_{ab}} = \frac{\sigma_{ao}}{r_{ao}} - \frac{\sigma_{bo}}{r_{bo}} \end{bmatrix}$$

FIGURE 5. Continuous phases a and b, with spherical interphases, in medium O. At equilibrium, forces of magnitudes σ_{ao}, σ_{bo}, and σ_{ab}, tangent to the a-o, b-o and a-b interphases, respectively, will balance one another. From this balance of forces arise the adhesive relations and corresponding equilibrium configurations. (From Phillips, H. M., Equilibrium Measurements of Embryonic Cell Adhesiveness: Physical Formulation and Testing of the Differential Adhesion Hypothesis, Ph.D. thesis, Johns Hopkins University, Baltimore, Maryland, 1969. Reproduced with permission from the author.)

$$\sigma_{ao} = \frac{W_{aa}}{2} \qquad W_{aa} = 2 \cdot \sigma_{ao} \tag{2}$$

$$\sigma_{bo} = \frac{W_{bb}}{2} \qquad W_{bb} = 2 \cdot \sigma_{bo} \tag{3}$$

and

$$\sigma_{ab} = \frac{W_{aa} + W_{bb}}{2} - W_{ab} \tag{4}$$

Finally,

$$W_{ab} = \sigma_{ao} + \sigma_{bo} - \sigma_{ab} \tag{5}$$

Figure 7 depicts a scheme to show how varying values of W_{ab} determine the configuration of a two-cell type aggregate after sorting out.

Dolowy[18] has discussed the physical theory of cell interactions and examined the rela-

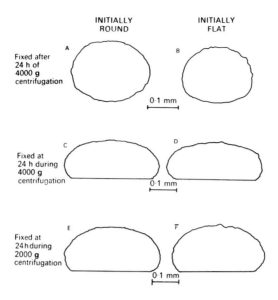

FIGURE 6.   Profiles of centrifuged aggregates traced from photographs. Note that the shape assumed after centrifugation depends on the tissue and is irrespective of the initial shape. A and B, chick limb bud; C and D, chick heart; E and F, chick liver. (From Phillips, H. M. and Steinberg, M. S., *J. Cell Sci.*, 30, 1, 1978. With permission.)

tionship between the cell surface free energy and sorting out behavior. Sorting out can occur in a mixture of randomly aggregated cells as in the experimental model or as an interaction between two layers of cells with different surface properties. The latter is specially interesting, since it is more akin to in vivo situations. The cell's energy of adhesion can determine the kind of interactions as summarized in Table 4, which is taken from the article of Dolowy[18] with a change in the notation for the surface energy. Dolowy[18] represents the energy as E with subscripts to denote the cell type. Here, following Steinberg,[9] we use σ. Some of the cases of interaction energies given in Table 4 probably correspond to some familiar examples of interacting tissues in embryos. The example of $\sigma_{aa} < \sigma_{ab} < \sigma_{bb} < 0$ is reminiscent of avian epiblast cells penetrating into the lower layer at the primitive streak. Separation of two layers (such as those of the lateral plate mesoderm) may be due to the energy relation represented by $\sigma_{aa} < \sigma_{bb} < 0 < \sigma_{ab}$. It is worth attempting a verification of these postulates by determining the surface energies of the cells involved in actual morphogenetic interactions.

The model presented by Steinberg is capable of accounting for cell sorting in a satisfactory manner, provided that its basic postulates are proved experimentally. In particular, the suggestion that the tissue hierarchy observed in sorting out arises from the graded difference in the value of σ for different tissues needs to be proved. Phillips and Steinberg[16,19] have obtained experimental evidence to show that chick limb bud precartilage, heart ventricle, and liver have decreasing values of σ in the given order, and this corresponds to their hierarchy in sorting out. This is, however, not quite satisfactory since two of the tissues, viz., heart and liver, exhibit what is called reversal in positioning. The evidence has therefore to be supplemented by more information on the σ of other tissue types depicted in the tissue hierarchy. The surface free energies of cells can determine their adhesive strength. Attempts to verify this hypothesis with reference to cell adhesive strength have failed to yield unequivocal conclusions. Using the method of collecting aggregate to determine the adhesiveness of cells, Roth and Weston[20] and Roth[21] could not find support for the hypothesis. More recently, Grover et al.[7] determined the adhesiveness of cells involved in nine binary com-

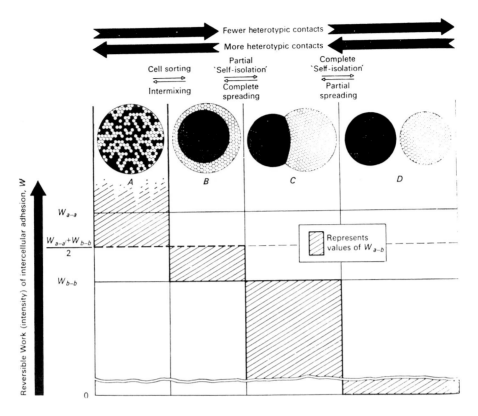

FIGURE 7.   Diagram to illustrate how the reversible works of cohesion $(W_{a-a}$ and $W_{b-b})$ and adhesion (W_{a-b}) determine the most stable configuration of a liquid system. These relationships should apply to any multi-subunit system that adopts liquid-like equilibrium shapes, whether the subunits are molecules or cells. (From Steinberg, M. S., Cell-cell recognition in multicellular assembly: levels of specificity, *Symp. Soc. Exp. Biol.*, 32, 25, 1978. With permission of the publishers, the Company of Biologists, Cambridge University Press.)

binations. The evidence was contrary to the hypothesis in four cases; in the rest, the differences in adhesiveness were considered to be not significant statistically.

It has been pointed out by Steinberg[17] that the various methods used for a "direct" measure of cell adhesiveness probably do not measure the property that is relevant to cell sorting. The "direct" methods determine the avidity of adhesion during chance collisions, whereas the cell property relevant to sorting out is related to the changing of already existing adhesive contacts to achieve a state of minimum free energy. In view of this, the evidence cited above cannot be considered as disproving the hypothesis. Overton[22] has shown that structural changes including the formation of desmosomes is more relevant to sorting out. According to her, the enclosed cells establish such junctions more frequently than the enclosing ones. This may be considered as an independent support to the hypothesis. More recently, Nicol and Garrod[23] have determined the mutual adhesive strength of the cells in the following order: corneal epithelium ≈ liver parenchyma > pigmented epithelium > limb bud mesenchyme. This roughly reflects their hierarchy in sorting out. Jones[24] has recently discussed the problem and emphasized that the formation of junctional complexes regulates cell positioning in sorting out. Nevertheless it is desirable to obtain more quantitative data on the different tissues to see if this property reflects the hierarchy.

B. Cell Surface Charge and Sorting Out

The cell surface electrical properties can presumably determine the kinds of associations

Table 4
CELL-CELL INTERACTION ENERGIES, CELL SORTING OUT, AND CELL POSITIONING

Interaction energies	Mixture of *a* and *b* cells	Layer of *a* cells covered with *b* cell layers
$\sigma_{ab} < \sigma_{a\text{-}a} < \sigma_{b\text{-}b} < 0$	No sorting out	Both cell types penetrate the opposite layer
$\sigma_{aa} < \sigma_{ab} < \sigma_{b\text{-}b} < 0$[a]	No sorting out	*b* Cells penetrate *a* layer
$\sigma_{ab} < \sigma_{a\text{-}a} < 0 < \sigma_{b\text{-}b}$	No sorting out; *b* cells do not adhere to themselves	*b* Cells penetrate *a* layer
$\sigma_{a\text{-}a} < \sigma_{a\text{-}b} < \sigma_{b\text{-}b} < 0$[b]	Aggregate of *a* cells within *b* aggregate	Stable
$\sigma_{a\text{-}a} < \sigma_{b\text{-}b} < \sigma_{ab} < 0$	*a* and *b* aggregates in contact	Stable
$\sigma_{a\text{-}a} < \sigma_{b\text{-}b} < 0 < \sigma_{ab}$	Two separate aggregates	Separation of the layers
$\sigma_{a\text{-}a} < \sigma_{ab} < 0 < \sigma_{b\text{-}b}$	A few *b* cells within *a* cells aggregate rest in suspension	*b* Cell layer detaches, cells form a suspension
$\sigma_{a\text{-}a} < 0 < \sigma_{ab}$ or $\sigma_{b\text{-}b}$	Aggregate of *a* cells and suspension of *b* cells	*b* Cell layer detaches and cells form a suspension
$\sigma_{ab} < 0 < \sigma_{a\text{-}a}$ or $\sigma_{b\text{-}b}$	Adhesion only between unlike cells	Cells in suspension; small aggregates of unlike cells form
$0 < \sigma_{a\text{-}a}$, σ_{ab} or $\sigma_{b\text{-}b}$	All cells in suspension	All cells in suspension

[a] $\sigma_{a\text{-}a} + \sigma_{b\text{-}b} < 2\sigma_{ab}$
[b] $\sigma_{a\text{-}a} + \sigma_{b\text{-}b} > 2\sigma_{ab}$

Modified after Dolowy, K., *Cell Adhesion and Motility,* Curtis, A. S. G. and Pitts, J. D., Eds., Cambridge University Press, London, 1980, 39.

Table 5
CELL SURFACE CHARGE AND SORTING OUT

Observation	No. of combinations
1. pI Of enclosing cells > pI of enclosed cells	31
2. pI Difference not significant statistically	3
3. pI Of enclosing cells < pI of enclosed cells	4
4. Reversal in positioning[a]	2
Total	**40**

[a] These combinations do not give reproducible results: the same cell type may be found in the enclosing or enclosed phase.

From Rao, K. V., Grover, A., and Beohar, P. C., *Prog. Clin. Biol. Res.,* 151, 345, 1984. With permission.

between cells. An attempt has been made to see if the surface negative charge density of the cells is in any way related to the tissue hierarchy. It has been found that in general, the enclosing cell type has a higher pI compared with that of the enclosed cells. In 31 out of 40 binary combinations studied, the pI of enclosing cells was found to be significantly higher than that of the enclosed cells, whereas in 4 combinations the enclosed cells were of a higher pI value (see Table 5). Figure 8 represents the tissue hierarchy inferred from histological observations of cell aggregates after sorting out along with the pI values of the respective cell types. In general, the pI of the enclosing cell type is greater than that of the enclosed.

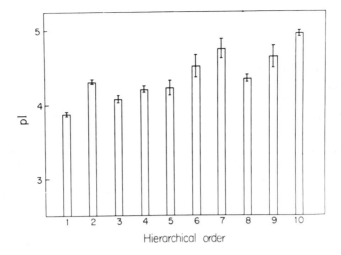

FIGURE 8. Cell pI arranged in the order of tissue hierarchy inferred from histological examination of cell sorting. The numbers refer to the tissues listed in Table 3. The vertical bars represent pI values with 95% confidence limits indicated. A gradation of pI values correlated with the hierarchical order may be noted. The correlation is statistically significant. (From Grover, A. and Rao, K. V., *Cell Differentiation*, 13, 209, 1983. With permission.)

Table 6
CELL pI AND ITS MODIFICATION BY HEPARIN

No.	Tissue type	pI[a] Control	Heparin treated
1.	Chick embryonic liver	4.52 ± 0.15	4.35 ± 0.10
2.	Chick embryonic neural retina	4.64 ± 0.15	4.30 ± 0.16
3.	Chick embryo leg muscle	4.23 ± 0.10	4.06 ± 0.05
4.	Chick embryonic brain	4.36 ± 0.05	3.82 ± 0.08
5.	Chick embryonic kidney	4.31 ± 0.03	4.36 ± 0.13
6.	Chick embryonic heart	4.21 ± 0.04	3.80 ± 0.10
7.	Rat embryonic brain	4.95 ± 0.04	3.85 ± 0.03
8.	Chick embryonic pigmented retina	3.88 ± 0.03	
9.	Chick embryo fibroblasts grown in vitro	4.08 ± 0.05	
10.	Rat embryonic kidney	4.75 ± 0.13	

[a] Mean pI values ± 95% confidence limits.

From Grover, A. and Rao, K. V., *Cell Differentiation*, 13, 209, 1983. With permission.

In order to test whether the apparent relationship between the hierarchical order and the pI values is significant, statistical analysis applicable to ranked variables was done. The correlation is statistically significant.[10]

A simple way of testing whether the apparent relationship between the cells' pI and the hierarchy is significant or just fortuitous is to modify the pI of cells and see the effect on cell sorting. It has been observed that generally the pI of cells decreases when they are treated with heparin (Table 6). The heparin molecules seem to bind on the surface and add negative charges.[25,26] The opportunity to modify the surface charge by heparin treatment was used in answering the question posed above. In many binary combinations in which one of the cell types was heparin treated, the position assumed by the cells was different from the corresponding controls. Rat embryonic brain cells when combined with a variety of other cells form an enclosing cortex, the other cell type forming the enclosed phase, respectively. Heparin-treated rat embryonic brain cells, on the contrary, constitute the enclosed phase when combined with these cells (Figure 4; Table 7).

Two tentative explanations can be offered to account for the change in the morphogenetic behavior of heparin treated cells: (1) after heparin treatment, the cells become passive or get altered in some nonspecific manner and are rendered unable to sort out properly, thus getting enclosed by the untreated cells that can move actively; (2) it is the altered pI of the treated cells that is responsible for their altered behavior. In case explanation (1) is right, one would expect that any cell type treated with heparin will be enclosed no matter what its altered pI is in relation to that of the other cell type combined with it. On the other hand, if the alternative explanation is right, the configuration would depend on the pI differences only. One can, on this basis, predict that a cell type treated with heparin could still be the enclosing phase, provided that its altered pI is higher than that of the untreated cell type with which it is combined. Table 7 lists four combinations (numbers 11 to 14) in which heparin-treated cells constitute the enclosing phase. Here, though the pI of the treated cells is slightly lowered, it is still greater than that of the other cell type they enclose. Morphogenetic sorting out is possible even when both the cell types are treated (numbers 15 to 17, Table 7). The mutual position of the cell types is in conformity with the rule:

$$pI_{enclosing\ phase} > pI_{enclosed\ phase}$$

It seems that the differences in surface charge somehow guide the cells to their equilibrium position in sorting out. The alteration caused by heparin seems to be at the surface of the cells. This is inferred from the observation that if heparin-treated cells are trypsinized and then allowed to sort out, there is no alteration in the pattern obtained. Whether the heparin molecules bind a plasma membrane component or a peripheral substance (such as fibronectin) is not clear.

It is premature to formulate any detailed theory on the role of cell surface charge in the process of cell sorting. However, the general conclusion that cells are guided by some interaction involving the surface charge can be a working hypothesis. It has been shown that the net negative charge of the cells measured by their electrophoretic mobilities is related to functional activities, especially locomotion in vitro.[27] An expression of the interfacial free energy discussed in connection with Steinberg's hypothesis may be the surface negative charge and the pI is its measure.

C. Interaction Modulation Factors and Cell Sorting

Finally we can consider another hypothesis briefly. Curtis[28,29] has proposed that almost all cell types produce diffusible substances called *morphogens* or *interaction modulation factors*. An important property of these substances is that they diminish the adhesiveness of unlike cell types in a relatively unspecific manner. The cell type that produces the particular morphogen is not responsive to it. Gradients of these factors will be set up in cell populations so that a pattern can be generated in the distribution of cells. It is also proposed that fibroblasts do not produce or respond to any modulation factor. The concentration gradients of the two distinct factors in a random aggregate of two cell types are never expected to be identical due to difference in their diffusibility. Figure 9 is a diagram explaining cell sorting in a mixed aggregate. Using two combinations, viz., chick embryonic neural retina + liver, and neural retina + heart, Curtis has obtained experimental evidence to suggest that the mechanism depicted in Figure 9 may be responsible for segregation as well as determination of the relative position of the cell types. Thus Curtis[28,29] attempts to reconcile the specific

Table 7
THE EFFECT OF HEPARIN ON CELL pI AND SORTING OUT BEHAVIOR[a]

No.	Sorting out in control combinations[c]	pI Difference[b]	Sorting out in experimental combinations[c]	pI Difference[b]
1.	Rat embryonic brain + (chick embryonic brain)		(Rat embryonic brain) + chick embryonic brain	
2.	Rat embryonic brain + (chick embryonic liver)		(Rat embryonic brain) + chick embryonic liver	
3.	Rat embryonic brain + (chick embryonic neural retina)		(Rat embryonic brain) + chick embryonic neural retina	
4.	Rat embryonic brain + (chick embryonic muscle)		(Rat embryonic brain) + chick embryonic muscle	
5.	Rat embryonic brain + (chick embryonic heart)		(Rat embryonic brain) + chick embryonic heart	
6.	Rat embryonic brain + (chick embryonic fibroblasts)		(Rat embryonic brain) + chick embryonic fibroblasts	
7.	Chick embryonic neural retina + (chick embryonic liver)	NS	(Chick embryonic neural retina) + chick embryonic liver	NS
8.	Chick embryonic brain + (chick embryonic muscle)	NS	(Chick embryonic brain) + chick embryonic muscle	
9.	Chick embryonic brain + (chick embryonic heart)		(Chick embryonic brain) + chick embryonic heart	
10.	Chick embryonic brain + (chick embryonic fibroblasts)		(Chick embryonic brain) + chick embryonic fibroblasts	
11.	Chick embryonic liver + (chick embryonic fibroblasts)		Chick embryonic liver + (chick embryonic fibroblasts)	
12.	Chick embryonic neural retina + (chick embryonic heart)		Chick embryonic neural retina + (chick embryonic heart)	NS
13.	Chick embryonic neural retina + (chick embryonic muscle)		Chick embryonic neural retina + (chick embryonic muscle)	NS
14.	Chick embryonic liver + (chick embryonic muscle)		Chick embryonic liver + (chick embryonic muscle)	
15.	Chick embryonic neural retina + (chick embryonic muscle)		Chick embryonic neural retina + (chick embryonic muscle)	NS
16.	Chick embryonic neural retina + (chick embryonic heart)		Chick embryonic neural retina + (chick embryonic heart)	
17.	Rat embryonic brain + (chick embryonic neural retina)		Chick embryonic neural retina + (rat embryonic brain)	

[a] Enclosed cell type is in parentheses.

[b] The mean pI of enclosing cells is greater than that of the enclosed. Where the difference is not statistically significant ($p > 0.05$), it is indicated by NS.

[c] Cell types printed in *italics* are heparin treated.

From Grover, A. and Rao, K. V., *Cell Differentiation*, 13, 209, 1983. With permission.

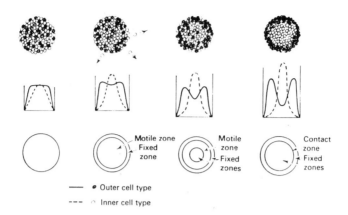

FIGURE 9.   A diagram to explain the sorting out of cell aggregates in terms of the interaction modulation theory. The top row shows a summary of the experimental observations during sorting out. The middle row shows how the two cell types establish gradients of their interaction modulation factors (IMFs) whose initially different slopes are determined by the different diffusibilities of the IMFs. These gradients then show positive feedback to the final forms as a result of the movement of cells to their sorted out positions. The bottom row shows main areas of cell movement during sorting out on this hypothesis. In any region where there is a considerable difference in the concentrations of the two IMFs, cells of the type opposite to that producing the greatest concentration of IMF will become less adhesive and will leave that region. (From Curtis, A. S. G., Cell-cell recognition: positioning and patterning system, *Symp. Soc. Exp. Biol.*, 32, 51, 1978. With permission of the publisher, the Company of Biologists, Cambridge University Press.)

adhesion and differential adhesion theories into a unified hypothesis. It is important to obtain more experimental evidence to verify the hypothesis. One would expect that some property of the morphogens should show a gradation reflecting the tissue hierarchy. Recently it has been found that BHK and HeLa cells do not sort out when mixed with chick embryonic cells.[30] It is likely that these cells grown in vitro as established cell lines are altered in such a manner as to be unable to produce the morphogens or to respond to them. An experimental verification of this has not yet been attempted.

III. SORTING OUT IN COMPLEX CELL AGGREGATES

An important goal of embryological studies is to discover the mechanisms that bring about the characteristic cellular displacements orchestrated to a finale defined by the anatomy of the new organism. Undoubtedly this is an immense task and is best attempted by resolving the highly complex process into simpler component processes for experimental analysis. Cell sorting as an experimental model has held out great hopes of discovering some of the mechanisms underlying morphogenetic processes. The research discussed in the preceding sections justifies continuing efforts in this direction. A closer examination of the available information shows that although the sorting out behavior of cells in most binary combinations is quite reproducible, there are a few in which the outcome is not constant. In other words, the same tissue may be the enclosed phase in relation to the other in some aggregates, while in others it may be the enclosing one. This feature is referred to as "reversal in tissue positioning". This has been found in combinations of chick embryonic heart with liver,[31] chick embryonic heart with limb precartilage or pigmented retina,[32] and chick brain with liver.[7]

Steinberg[9] attempted to explain these cases on the basis of cell adhesiveness and the tissue hierarchy. Tissues that are close in their hierarchical order probably do not differ in their adhesiveness (or any other property relevant to sorting out) and therefore may assume the enclosed or enclosing position. However, the "closeness" in hierarchical order is likely to change when sorting out involving more cell types is studied. Grover et al.[7] have "inserted" chick embryonic muscle, kidney, and brain between liver and heart in their hierarchical sequence. If Steinberg's explanation is valid, all combinations involving these tissues would be expected to exhibit reversal in cell positioning. However, this is not found to be so. Several workers have attempted to account for the reversal in positioning presumably in the hope that any explanation for these exceptions may turn out to be the clue to understanding the actual mechanism underlying the phenomenon of cell sorting.

Cell products, which can modify the developmental program of embryonic cells, have been called morphogens.[28,29] Investigating the effects of such morphogens may lead to a clue to reversal in tissue positioning. Wiseman et al.[31] reported that heart cell aggregates, which have been produced by reaggregation of dispersed cells or heart fragments, tested immediately following excision from the embryos tend to envelop liver, whereas heart fragments maintained in organ culture for 2 to 5 days prior to construction of fragment pairs tend to be enveloped. Later Armstrong[33,34] observed that the tissue affinities can be modulated by a cell surface protein produced by fibroblasts. Being smaller than most other cell types, the fibroblasts are probably not always included in the final aggregate during procedures of centrifugation. Culturing the aggregates or tissue fragments allows time for multiplication of the fibroblasts included in the cell preparation or tissue fragment. Chick embryonic heart is known to contain several cell types,[35] the most abundant of them being cardiac myocytes (70 to 80%) and fibroblasts (20 to 30%). Armstrong[33] studied the behavior of chick embryonic heart and pigmented retina in organ culture and found that the latter tissue envelops the former partially, in the absence of fibroblasts. When fibroblasts are present, the pigmented retina encloses the heart tissue. Further, Armstrong[34] studied the behavior of chick embryonic heart-heart aggregate pairs, both in the presence or absence of fibroblasts, and concluded that the phenomenon of reversal in positioning results from the activities of the fibroblasts. Armstrong[36] has presented further evidence to show that the substance that modulates the cell behavior is fibronectin.

The probable modulating effects of fibroblasts have been studied in a number of sorting out combinations. Fibroblasts grown in vitro (low passage, not more than three) were added as the additional third cell type to binary combinations of cells that had already been studied (Table 8).[37] This takes the experimental model through one step, approaching the complexity of actual embryonic organisms. The pattern obtained after sorting out is concentric, one tissue successively enclosed by the other two. The most striking feature of sorting out exhibited by the three-cell type combinations is the absence of any simple rule or pattern, which can be extrapolated from the study of binary combinations. In case the tissue hierarchy is the only guiding mechanism in the rearrangement of cells, one would expect to find the same hierarchical arrangement in the three-cell type combinations after sorting out. More than 40 such combinations have been studied histologically.[37] About a half of them reflect the hierarchy after sorting out. However the exceptions, though at present without an explanation, are too numerous to be ignored. The correlation of surface charge and positioning is also not evident in a large number of the cases. Clearly, therefore, in the aggregates of higher degrees of complexity, tissue hierarchy or a single surface property is not the only mechanism guiding the cells to a position of equilibrium.

Another way of looking at the available information is to consider that the fibroblasts do not exert any influence modulating the behavior of the other cell types in sorting out, as suggested by the hypothesis of Curtis.[28,29] In such a view, one may expunge fibroblasts from the list of cell types depicting the hierarchy. Further, one would expect that the fibroblasts will not influence the relative position of the other two cell types in any three-cell type combination. It is interesting to note that in a large number of the three-cell type combinations, fibroblasts do not disturb the hierarchical order of positioning of the other two cell types relative to each other. On balance however, it seems judicious not to generalize on the basis of the information available so far. It is imperative to know if other three-cell type com-

Table 8
SORTING OUT BEHAVIOR OF THREE-CELL TYPE COMBINATIONS

No.   **Position of cells after sorting out**[a]

1. Chick embryonic brain + [chick embryonic fibroblasts + (chick embryonic kidney)]
2. Chick embryonic neural retina + [chick embryonic fibroblasts + (chick embryonic kidney)]
3. Chick embryonic brain + [chick embryonic fibroblasts + (chick embryonic pigmented retina)]
4. Chick embryonic neural retina + [chick embryonic fibroblasts + (chick embryonic pigmented retina)]
5. Chick embryonic neural retina + [chick embryonic fibroblasts + (chick embryonic liver)]
6. Chick embryonic brain + [chick embryonic fibroblasts + (chick embryonic liver)]
7. Chick embryonic fibroblasts + [chick embryonic liver + (chick embryonic kidney)]
8. Rat embryonic kidney + [chick embryonic fibroblasts + (chick embryonic liver)]
9. Chick embryonic fibroblasts + [chick embryonic muscle + (chick embryonic liver)]
10. Chick embryonic fibroblasts + [chick embryonic heart + (chick embryonic liver)]
11. Chick embryonic brain + [chick embryonic fibroblasts + (chick embryonic muscle)]
12. Chick embryonic fibroblasts + [chick embryonic muscle + (chick embryonic kidney)]
13. Chick embryonic neural retina + [chick embryonic fibroblasts + (chick embryonic muscle)]
14. Rat embryonic kidney + [chick embryonic fibroblasts + (chick embryonic muscle)]
15. Chick embryonic brain + [chick embryonic fibroblasts + (chick embryonic heart)]
16. Chick embryonic fibroblasts + [chick embryonic heart + (chick embryonic kidney)]
17. Chick embryonic fibroblasts + [chick embryonic heart + (chick embryonic pigmented retina)]
18. Chick embryonic brain + [chick embryonic fibroblasts + (rat embryonic kidney)]
19. Chick embryonic neural retina + [rat embryonic fibroblasts + (chick embryonic liver)]
20. Chick embryonic neural retina + [rat embryonic fibroblasts + (chick embryonic kidney)]
21. Chick embryonic neural retina + [rat embryonic fibroblasts + (chick embryonic muscle)]
22. Chick embryonic neural retina + [rat embryonic fibroblasts + (chick embryonic heart)]
23. Chick embryonic brain + [rat embryonic fibroblasts + (chick embryonic kidney)]
24. Chick embryonic brain + [rat embryonic fibroblasts + (chick embryonic liver)]
25. Chick embryonic brain + [rat embryonic fibroblasts + (chick embryonic muscle)]
26. Chick embryonic brain + [rat embryonic fibroblasts + (chick embryonic heart)]
27. Rat embryonic fibroblasts + [chick embryonic muscle + (chick embryonic liver)]
28. Rat embryonic fibroblasts + [chick embryonic liver + (chick embryonic kidney)]
29. Rat embryonic fibroblasts + [chick embryonic muscle + (chick embryonic kidney)]

[a]   [   ] Enclosed cell types. (   ) Cell type forming the innermost phase.

Courtesy of Grover, A., Sharma, S., and Rao, K. V., Unpublished data.

binations (i.e., those without fibroblasts as a component) show the hierarchical order after sorting out. In a preliminary study, Sharma and Rao[38] found no hierarchical order in 10 of the 15 such combinations. The conclusion is that we ought to have more information and also consider the possibility of other factors guiding the cells in sorting out.

IV. HISTOGENESIS IN THE EXPERIMENTAL MODELS

We have noted earlier that the disaggregated cells organize themselves often into histologically recognizable patterns after sorting out (Figure 3). We have also noted that the homotypic cellular zones after sorting out do not organize themselves into anatomically normal arrangement, even when the tissue types are related anatomically. Thus when neural and pigmented retinal cells are allowed to sort out, the resulting arrangement is totally unlike what is found in an embryo. From this it is evident that cell sorting as a model of morphogenesis has only a limited value in elucidating embryogenesis. In order to derive further benefits from the study of this experimental model, cell combinations have to be selected more judiciously, and the criteria of assessing the results have to be defined freshly. A simple and grossly manifest difference like the enclosed and enclosing position obviously does not reflect the subtle morphogenetic rearrangements occurring in embryonic tissues.

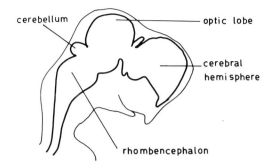

FIGURE 10.   Diagram showing the anatomy of the brain of an early chick embryo.

A step forward in further exploiting the experimental model has been taken in the interesting work reported by Spiegel and Garber.[39] Regional differentiation and the establishment of cell connections in the embryo seem to depend on region-specific cell recognition molecules. The vertebrate central nervous system is composed of distinct regions in which cells interact by establishment of specific associative connections. The cerebrum and optic tectum (dorsal wall of the optic lobes, see Figure 10) are distinct regions of the brain. Spiegel and Garber[39,40] obtained the cerebrum-specific and optic tectum-specific factors by culturing the respective chick embryonic cells in a serum free medium. The "factors" (cerebrum-specific factor [CRF] and optic tectum specific factor, [OTF]) were shown to have the property of enhancing the aggregate size of cells of the respective brain regions. Stage specificity is also exhibited in this property. Cells of a day older or younger embryos did not respond to the factor. However the factors could act beyond broad taxonomic barriers. The CRF could enhance the size of aggregates formed by quail or mouse cerebrum of comparable stage of development.

When the cells of cerebrum and optic tectum were mixed randomly (coaggregated in agitated suspensions), they did not sort out into the familiar enclosed and enclosing phases. However, under the influence of the region-specific cell recognition factors, the cells in the aggregates started establishing homotypic associations. The homologous and heterologous interactions of cell aggregates from chick and quail embryos were determined by the nearest neighbor analysis estimating what is called the α-cross ratio. If the two cell types are distributed randomly, their α-cross ratio will be 1.0. As the cell types sort out into homotypic zones, the ratio will approach 0. Thus the α-cross ratio is an index of cell sorting within a mixed aggregate. As shown by the data of Spiegel and Garber,[39,40] the addition of factors enhances the tendency of the cells to establish homotypic associations. This was also concluded from a dose-response relationship between the factor and the respective cell associations.

Investigations using cell sorting as an experimental model are still continuing. Information, which can be extrapolated to embryonic morphogenesis, has yet to emerge from the experimental model. However, it is abundantly clear that the model offers unique opportunities for analysis of the morphogenetic behavior of cells. The recent work on the modulation of morphogenetic behavior of tissue cells by fibroblasts and their extracellular products should be pursued further. The ubiquitous fibroblasts, which are known to be heterogeneous with respect to surface properties,[41] may well be an important factor underlying the mechanism of cellular activities leading to the formation of embryonic tissues.

REFERENCES

1. **Townes, P. L. and Holtfreter, J.,** Directed movements and selective adhesion of embryonic amphibian cells, *J. Exp. Zool.,* 128, 53, 1955.
2. **Moscona, A. A.,** Rotation-mediated histogenetic aggregation of dissociated cells, *Exp. Cell Res.,* 22, 455, 1961.
3. **Moscona, A. A.,** Analysis of cell recombinations in experimental synthesis of tissues *in vitro, J. Cell Comp. Physiol.,* 60(Suppl. 1), 65, 1962.
4. **Burdick, M. L. and Steinberg, M. S.,** Embryonic cell adhesiveness: do species differences exist among warm blooded vertebrates?, *Proc. Natl. Acad. Sci. U.S.A.,* 63, 1169, 1969.
5. **Garber, B. B. and Moscona, A. A.,** Aggregation *in vivo* of dissociated cells. I. Reconstruction of skin in the chorioallantoic membrane from suspensions of embryonic chick and mouse skin cells, *J. Exp. Zool.,* 155, 179, 1964.
6. **Grady, S. R. and McGuire, E. J.,** Intercellular adhesive selectivity. III. Species selectivity of embryonic liver intercellular adhesions, *J. Cell Biol.,* 71, 96, 1976.
7. **Grover, R., Grover, A., and Rao, K. V.,** Experimental analysis of morphogenetic cell sorting, *Indian J. Exp. Biol.,* 18, 1072, 1980.
8. **Nag, A. C., Cheng, M., and Healy, J. C.,** Coaggregation and formation of a joint myocardial tissue by embryonic mammalian and avian heart cells, *J. Embryol. Exp. Morphol.,* 59, 263, 1980.
9. **Steinberg, M. S.,** Does differential adhesion govern cell assembly process in histogenesis? Equilibrium configurations and the emergence of a hierarchy among populations of embryonic cells, *J. Exp. Zool.,* 173, 395, 1970.
10. **Grover, A. and Rao, K. V.,** Sorting out in heterotypic cell aggregates is regulated by differences in the cell surface charge, *Cell Differentiation,* 13, 209, 1983.
11. **Lilien, J. E.,** Specific enhancement of cell aggregation *in vitro, Dev. Biol.,* 17, 657, 1968.
12. **Edelman, G. M.,** Cell adhesion molecules, *Science,* 219, 450, 1983.
13. **Kuroda, Y.,** Preparation of an aggregation-promoting supernatant from embryonic chick liver cells, *Exp. Cell Res.,* 49, 626, 1968.
14. **Garber, B. B. and Moscona, A. A.,** Reconstruction of brain tissue from cell suspensions. II. Specific enhancement of aggregation of embryonic cerebral cells by supernatant from homologous cell cultures, *Dev. Biol.,* 27, 235, 1972.
15. **Steinberg, M. S.,** On the mechanism of tissue reconstruction by dissociated cells. III. Free energy relations and the recognition of fused heteronomic tissue fragments, *Proc. Natl. Acad. Sci. U.S.A.,* 48, 1769, 1962.
16. **Phillips, H. M. and Steinberg, M. S.,** Embryonic tissues as elasticoviscous liquids. I. Rapid and slow shape changes in centrifuged cell aggregates, *J. Cell Sci.,* 30, 1, 1978.
17. **Steinberg, M. S.,** Cell-cell recognition in multicellular assembly: levels of specificity, *Symp. Soc. Exp. Biol.,* 32, 25, 1978.
18. **Dolowy, K.,** A physical theory of cell-cell and cell-substratum interactions, in *Cell Adhesion and Motility,* Curtis, A. S. G. and Pitts, J. D., Eds., Cambridge University Press, London, 1980, 39.
19. **Phillips, H. M. and Steinberg, M. S.,** Equilibrium measurements of embryonic cell adhesiveness. I. Shape equilibrium in centrifugal fields, *Proc. Natl. Acad. Sci. U.S.A.,* 64, 121, 1969.
20. **Roth, S. and Weston, J. A.,** The measurement of intercellular adhesion, *Proc. Natl. Acad. Sci. U.S.A.,* 58, 974, 1967.
21. **Roth, S.,** Studies on intercellular adhesive selectivity, *Dev. Biol.,* 18, 602, 1968.
22. **Overton, J.,** Responses of epithelium and mesenchymal cells to culture on basement lamella observed by scanning electron microscopy, *Exp. Cell Res.,* 105, 313, 1977.
23. **Nicol, A., and Garrod, D. R.,** Fibronectin, intercellular junctions and the sorting out of chick embryonic tissue cells in monolayer, *J. Cell Sci.,* 54, 357, 1982.
24. **Jones, B. M.,** Aspects of cell sorting in aggregates, *Prog. Clin. Biol. Res.,* 151, 275, 1984.
25. **Chaubal, K. A. and Lalwani, N. D.,** Electrical capacity of external cell surface: electrophoretic mobility analysis with polyanion treatment, *Indian J. Biochem. Biophys.,* 14, 285, 1977.
26. **Rao, K. V., Grover, R., and Mehta, A.,** Isoelectric focusing of cells using zwitterionic buffers, *Exp. Cell Biol.,* 47, 360, 1979.
27. **Bhisey, A. N., Rao, G. S. A., and Ranadive, K. J.,** Alterations in electrophoretic mobility of mouse macrophages treated with microtubule inhibitors, *Indian J. Exp. Biol.,* 15, 970, 1977.
28. **Curtis, A. S. G.,** Cell positioning, in *Specificity of Embryological Interactions,* Garrod, D. R., Ed., Chapman & Hall, London, 1978, 157.
29. **Curtis, A. S. G.,** Cell-cell recognition: positioning and patterning system, *Symp. Soc. Exp. Biol.,* 32, 51, 1978.
30. **Sharma, S., Rao, K. V., Joshi, M. V., Chiplonkar, J. M., and Wagh, U. V.,** Unpublished observations.

31. **Wiseman, L. L., Steinberg, M. S., and Phillips, H. M.,** Experimental modulation of intercellular cohesiveness: reversal of tissue assembly patterns, *Dev. Biol.,* 28, 498, 1972.
32. **Armstrong, P. B. and Niederman, R.,** Reversal of tissue positioning after cell sorting, *Dev. Biol.,* 28, 518, 1972.
33. **Armstrong, P. B.,** Modulation of tissue affinity of cardiac myocytes by mesenchyme, *Dev. Biol.,* 64, 60, 1978.
34. **Armstrong, P. B.,** Ability of cell surface protein produced by fibroblasts to modify tissue affinity behaviour of cardiac myocytes, *J. Cell Sci.,* 44, 263, 1980.
35. **DeHaan, R. L.,** Regulation of spontaneous activity and growth of embryonic chick heart cells in tissue culture, *Dev. Biol.,* 16, 216, 1967.
36. **Armstrong, P. B.,** Role of extracellular matrix in the control of cell motility in a model morphogenetic system, *Prog. Clin. Biol. Res.,* 151, 309, 1984.
37. **Grover, A., Sharma, S., and Rao, K. V.,** Unpublished observations.
38. **Sharma, S. and Rao, K. V.,** Unpublished observations.
39. **Spiegel, J. and Garber, B.,** Sorting out of coaggregated brain cell types mediated by specific cell recognition factors *in vitro.* I. Experimental histogenesis and quantitative analysis, *Dev. Biol.,* 85, 1, 1981.
40. **Spiegel, J. and Garber, B.,** Sorting out of coaggregated brain cell types mediated by specific cell recognition factors *in vitro.* II. Kinetic and metabolic parameters, *Dev. Biol.,* 85, 16, 1981.
41. **Garrett, D. M. and Conrad, G. W.,** Fibroblast-like cells from embryonic chick cornea, heart and skin are antigenically distinct, *Dev. Biol.,* 70, 50, 1979.
42. **Hamburger, V. and Hamilton, H. L.,** A series of normal stages of development of the chick embryo, *J. Morphol.,* 88, 49, 1951.
43. **Rao, K. V., Grover, A., and Beohar, P. C.,** Cell sorting: an experimental model to elucidate the cellular basis of morphogenesis, *Prog. Clin. Biol. Res.,* 151, 345, 1984.

Chapter 8

THE CELLULAR BASIS OF EMBRYOGENESIS

I. INTRODUCTION

In the previous chapters, we have discussed the cellular basis of simple morphogenetic changes in some model systems. The "developmental" changes undergone by the cellular slime molds, and the reconstituting sponge or vertebrate embryonic tissue cell aggregates are undoubtedly spectacular. However, the pageant exhibited by intact animal embryos is still more fascinating. The aim of developmental biology is to unravel the mechanisms underlying embryonic development. In a sense, experimental models such as the cellular slime molds serve as easy preliminary exercises preparing us to tackle the more difficult and challenging problems of elucidating the development of embryos.

Embryonic development starts from the time of fertilization. A single cell, viz., the fertilized egg, undergoes repeated divisions, which exhibit a definite pattern in space and time. Asynchrony is apparent in the cell divisions sooner or later, depending on the distribution of yolk and possibly owing to other characteristics. Often the size of the resulting blastomeres is clearly unequal. The planes of cleavage in many animal eggs are known to bear a definite relation with the future morphological organization of the embryo. Experimental embryologists have demonstrated that in some animals, parts of the embryo can already be identified and demarcated on different blastomeres, thus tracing cell lineages. Yet in many cases, experimental rearrangement of the cells or even the elimination of some of them is without deleterious effects. These features are precisely controlled by mechanisms largely inherent in the fertilized egg. An inquiry into the nature of these mechanisms is a continuing effort undertaken by generations of embryologists.

It is almost a truism that all developmental events are regulated directly or indirectly by the surface properties of the cells involved. The differences, demonstrated or suspected, in the cell surface properties (e.g., lectin agglutinability, presence or absence of a certain antigen, receptors of external signals, etc.) certainly suggest that they could be of developmental significance. However, no precise knowledge exists of how these differences control a given process of development. In this chapter, we shall discuss some of the embryonic developmental processes in which considerable work has been done, demonstrating or suggesting a definite role for changes in the cell surface. The diversity of embryonic forms and developmental events necessitates choosing illustrative examples. Much of the fundamental work has been carried out on vertebrate embryos. In view of this, most of the examples have been selected from among them.

II. EARLY DEVELOPMENT: THE DETERMINATION OF EMBRYONIC AXES AND CLEAVAGE

A. The Embryonic Axes

One of the earliest events in the development of an animal embryo is the determination of the antero-posterior and dorso-ventral axes. In many animal eggs, this is not determined before fertilization. There is, however, a distinct polarity in the unfertilized egg represented by the animal-vegetal pole axis. Generally, the yolk deposited in the egg has a definite relation with the animal-vegetal axis. This is especially evident in the telolecithal eggs which show a greater concentration of the yolk towards the vegetal pole. Unequal distribution of yolk is not the only component of this axial organization. In the marine scaphopod *Dentalium* and many gastropods, there is a specialized vegetal pole cytoplasm known as the polar lobe, which has distinct morphogenetic properties.[1] There are other examples of some special material located at the vegetal pole. The germ plasm of *Xenopus* has been demonstrated to be located at the vegetal pole.[2] Lateral mobility of plasma membrane lipids shows regional differences related to the animal-vegetal polarity. Following fertilization, these differences are accentuated.[3] In many insect eggs, the posterior pole of the egg has the germ plasm as well as other cytoplasmic structures that are essential for the development of germ cells.[4] How the animal-vegetal polarity originates is not known. It is possible that it owes its origin to the distribution of the sites of vitellogenin pinocytosis on the surface of the growing oocyte. Where receptors are more densely distributed, the yolk would be more abundantly located.[5] There is, however, no direct evidence to support this hypothesis. Before fertilization, the egg seems to be radially symmetrical around the animal vegetal axis. The point of entry of the sperm defines a new point on the egg surface, thus making one of the meridians distinguished from the others. Since the entry of the sperm involves considerable membrane alterations, it is probable that it introduces some changes in the organization of the egg cytoplasm, and consequently the future development could be regulated by it.

Early embryologists have recognized two important events that are related to the determination of the dorso-ventral and antero-posterior axes of the amphibian egg: (1) the sperm enters randomly at one point in the animal hemisphere defining the plane of the first cleavage and that of bilateral symmetry of the future embryo; (2) the fertilized uncleaved egg develops a crescent shaped area, the grey crescent, which marks the part of the egg that eventually gives rise to the dorsal structures of the embryo. In other words, the primary axial organization of the future embryo is established at the time of fertilization. Vital dye marking and other experiments demonstrated that the blastopore appears at the lower limit of the grey crescent. Since a topographical relationship exists between the grey crescent and the axial organization of the embryo, considerable interest has been shown by many workers in demonstrating a causal relation between the two. Some of the important reviews on the grey crescent problem are Ancel and Vintemberger,[6] Pasteels,[7] Nieuwkoop,[8] and Brachet.[5] Though the grey crescent is manifested by changes in the intensity of pigment in the region, it does not seem to be related to the pigment per se. Rather, the pigment is only a visible marker of some change that occurs in the egg cortex, i.e., the cytoplasm immediately beneath the plasma membrane. In fact, in poorly pigmented or pigmentless eggs, there is no visible equivalent area that manifests itself at the homologous site. Yet, functionally it exists as an embryological entity.

According to Ancel and Vintemberger,[6] the cortex of the egg rotates 30° around the axis perpendicular to the plane of symmetry resulting in stretching of the cortex on the future dorsal side (Figure 1). This is known as the cortical reaction of symmetrization. At the same time, under the influence of gravity, the yolk undergoes displacement. As a result, the cortex of the future dorsal side comes closer to the large platelets of the vegetal yolk. Experimentally it has been demonstrated that the newly acquired closer contact between the dorsal egg cortex and the vegetal yolk somehow determines the bilateral symmetry definitively. It has been suggested that there is an interaction between the cortical region and the yolk. Since the yolk is distributed differentially as an animal vegetal gradient, it has a vectorial property. Similarly the cortex is also assumed to show a gradient. It is suggested that the product of the two gradients is also a gradient, and the highest value of the combined gradient determines the dorsal region and the eventual center of invagination during gastrulation. Pasteels[7] demonstrated the influence of the "cortical field" and "yolk field" by displacing the vegetal yolk in relation to the dorsal cortex. Axolotl eggs freed from the vitelline membrane were placed in a stiff gel which prevented their normal free return under gravity when they were turned. Eggs were inverted and placed on the stiff gel as shown in Figure 2. It was observed that the blastoporal lip always appeared at the margin of vegetal yolk nearest the center of the grey crescent. The cortical field is unaffected by gravity, whereas the yolk distributes itself when rotated. Pasteels[7] explained these observations on the basis of the cortical field hypothesis.

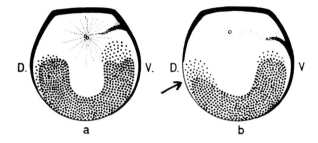

FIGURE 1.   Schematic drawings of vertical sections through eggs of *Rana temporaria*, just before (a), and immediately after (b), the reaction of symmetrization. The cortical pigment is shown in black, vegetal yolk heavily dotted and the grey crescent shown by the arrow; dorsal side, D, ventral side, V. (From Ancel, P. and Vintemberger, P., *Bull. Biol. Suppl.*, 31, 1, 1948, as adapted by Pasteels, J. J. in *Adv. Morphogenesis*, 3, 363, 1964. With permission.)

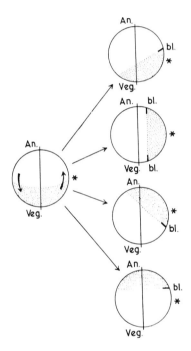

FIGURE 2.   Schematic drawings of eggs of axolotl that have been partially turned, but at different angles, around an axis perpendicular to symmetry. Left, before the experiment; right, after the experiment. An, animal pole; Veg., vegetal pole; bl, dorsal lip of the blastopore; *, center of the grey crescent; dotted area, heavy yolk. (From Pasteels, J. J., *Adv. Morphogenesis*, 3, 363, 1964. With permission.)

Recently Gerhart et al.[9] have repeated and extended the investigation using another amphibian species, viz., *Xenopus laevis*. When a fertilized egg with the vitelline membrane intact is inverted, it reestablishes the vertical animal-vegetal axis due to gravity. The perivitelline space contains a fluid which permits free rotation of the egg within it. When the egg is immersed in a solution of 5% Ficoll (a sucrose polymer), the perivitelline fluid is dehydrated and the vitelline membrane collapses over the egg surface preventing free rotation of the egg under gravity. In their experiments involving rotation of the egg, Gerhart et al.[9]

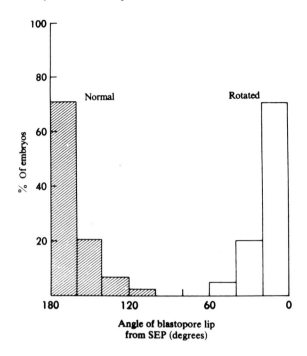

FIGURE 3.   The topographical relationship of the point of entry of the
sperm (SEP) to the location of the dorsal blastoporal lip in *Xenopus laevis*
embryos. See text. (Reprinted by permission from *Nature (London)*, Vol.
292, No. 5823, p. 512. Copyright, 1981, Macmillan Journals Ltd.)

used this method. An advantage of this method is that the perivitelline fluid can be rehydrated
at will and the egg permitted to rotate within the space.

In the majority of the normal eggs of *Xenopus,* the blastoporal lip appears 135 to 180°
away from the point of entry of the sperm (Figure 3). This could be demonstrated by placing
a vital dye (Nile blue) spot at the equator on the meridian defined by the point of entry of
the sperm and observing the blastoporal lip, which develops eventually, with reference to
the dye spot. In experiments involving inversion of the egg, a vital dye mark was placed
just below the point of entry of the sperm, at the equator of the egg. The egg was then
rotated so that the dye mark is pointing upwards and the animal-vegetal axis, horizontal
(Figure 4). After the egg had been retained in this condition for varying durations, it was
allowed to assume the position under gravity by rehydrating the perivitelline fluid. The result
was striking: in 70% of the eggs thus treated, the blastopore developed from 0 to 20° from
the mark; in 20%, in the sector 20 to 40° away from it and in 10% within 40 to 60° (see
Figures 3 and 4). When the egg was inverted with the dye mark pointing downward, the
blastopore appeared at the point where the grey crescent formed. In other experiments, the
rotation was to the right or left rather than the top. In all cases, the point of blastopore
formation corresponded to the point on the equator uppermost during the period of rotation.
In the postcrescent period, the egg became more and more refractory to the effect of gravity.
Centrifugation could enhance the effect of gravity. Gerhart et al.[9] concluded that the entry
of the sperm followed by the formation of a large aster brings about a localized contraction
of the egg cortex on the opposite side and also causes displacement of the cytoplasmic
contents leading to the establishment of dorso-ventral axis. It appears that it is the cytoplasmic rearrangement rather than the grey crescent per se that determines the dorso-ventral
axis. An additional factor, which has a functional role in this process, is the cytoskeleton
of the egg.[10,11]

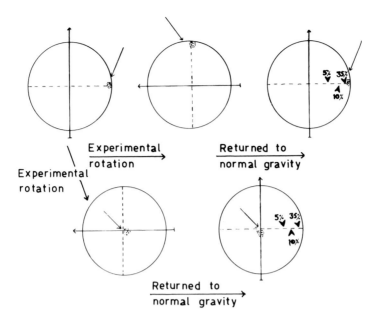

FIGURE 4. Diagram to show the experimental rotation of *Xenopus* eggs and the resulting axial determination. The long arrows represent the animal-vegetal axis of the egg. The short arrows show the point of entry of the sperm. The dotted line represents the equator of the egg. The arrow heads show the frequency of dorsal blastoporal lip formation with reference to the vital dye mark, shown dotted. (Based on Gerhart, J., Ubbels, G., Black, S., Hara, K., and Kirschner, M., *Nature (London)*, 292, 511, 1981.)

    The process of axis determination in avian embryos depends primarily on the influence of gravity. Axis determination occurs while the egg is still in the uterus. Kochav and Eyal-Giladi[12] showed that the position of the egg in the hen's uterus is such that the future posterior end is always developed at the blastodisc's highest point. The temporal limits of axis determination span between 14 and 16 hr from the time of release into the uterus. During this period, it is possible to manipulate the position of the egg and alter the orientation of the embryonic axis. Unlike in the amphibian embryos, axis determination in the chick seems to be a gradual process determined by the manner in which the hypoblast is formed as a separate layer. In fact, the influence of the hypoblast on the formation of the primitive streak has been known for a long time.[13] The work of Eyal-Giladi and colleagues,[14,15] which involves some extraordinary experimental skill, has now shown that the earliest events in the determination of the antero-posterior axis are influenced by gravity. Further stabilization of the axis depends on an interaction between the hypoblast and the epiblast. The influence of the hypoblast is in the nature of a gradient of inductivity, the physical basis of which is, however, obscure. Whether the gradient is due to any chemicals or a spatial distribution of the cell types is not known. It has been suggested[16] that little pockets beneath the hypoblast, which coalesce to form a bigger pocket, have a positive role to play in this. In view of the fact that axis determination is a fundamental developmental process and that embryonic development takes place apparently always under the influence of gravity, it may be concluded that the two are causally related. Obviously gravitational force stabilizes the distribution of a variety of materials in the embryo, and this may have far-reaching consequences to all subsequent development. Further, it seems reasonable to expect that there must be a common mechanism of axis determination, which has been conserved in large groups of animals such as the vertebrates. The ultimate test of gravity as a causal factor would be to fertilize eggs in outer space and study their development. Such an experiment using *Xenopus*

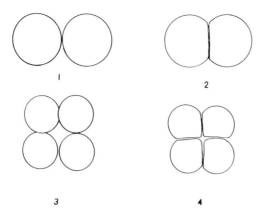

FIGURE 5.   Changing shape of the blastomeres during the early cleavage of *Lymnaea* egg.

eggs has been planned by Dr. G. Ubbels[17,18] of the Hubrecht Laboratory (Utrecht, The Netherlands). The outcome of this experiment, which is already under way, is eagerly awaited.

Axial organization is a decisive and critical event in the development of an embryo. Considerable information is available on the axial organization of ascidian and amphioxus embryos and its correlation with changes in the egg cytoplasm following fertilization.[19]

How are the cytoplasmic and cell surface events during fertilization and cleavage related to programing particular blastomeres for a certain pathway of differentiation among several possible ones? It may be suggested *a priori* that the initiation of invagination is controlled by the cytoskeletal machinery. Of course, there could be other causes too. But how could the very early events following fertilization control the organization of cytoskeleton and canalize particular cells to differentiate in such a manner that they have the right kind of surface specializations and to respond to external signals such as those of the extracellular materials? Nothing is known at present to link the events. Clearly, this is an area in which some vital information is missing.

B. Cleavage

The fertilized egg engages in a rapid process of cell divisions. The plane of cleavage and its pace in the different blastomeres in relation to each other are precisely controlled, as known from descriptive embryological studies. In eggs with scanty or moderate amounts of yolk, cleavage results in the division of the whole egg (holoblastic cleavage) whereas in heavily yolk-filled eggs, it is only partial (meroblastic cleavage) and restricted to the relatively yolk-free region. Excepting in the bizarre embryos of some Platyhelminthes, the cells resulting from the cleavage remain as a single mass. Evidently there is a definite adhesive mechanism holding the cells together. The degree of adhesive strength varies. In certain molluscs, the blastomeres are alternately round and flattened at the faces of mutual contact during each cycle of cell division (Figure 5). There must be a mechanism consisting of the plasma membrane and cytoskeletal elements controlling these changes. In all probability, there is some significance to this curious phenomenon which, however, is unknown.

Differences in the size of blastomeres resulting from cleavage are of significance to the subsequent developmental events. What mechanisms other than yolk content determine the size differences is unknown. That additional mechanisms must exist is, however, clear from many instances where blastomeres vastly differing in size are developed, though there is no corresponding disparity in the distribution of yolk. Eggs that exhibit the typical spiral cleavage are striking examples of this. During the third and subsequent cleavages, very small cells

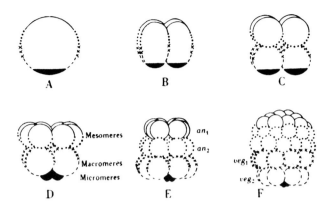

FIGURE 6. Early cleavage of the sea urchin egg: (A) uncleaved egg with different zones in the animal-vegetal axis shown by different markings that may be traced into the subsequent stages; (B) 4-cell stage, resulting from two vertical cleavages of the egg at right angles to each other; (C) 8-cell stage, resulting from a horizontal cleavage of the four cells; (D) 16-cell stage which results from a vertical cleavage of the animal half and a horizontal cleavage of the vegetal half. Note that the horizontal cleavage is very unequal, resulting in the formation of small micromeres (shown black). (E) and (F) Further cleavages of the blastomeres; an_1, an_2, veg_1, and veg_2 are different zones of the egg from the animal to the vegetal pole. When separated and grown in isolation, these zones show different patterns of differentiation. (After Hörstadius, S., *Biol. Rev.*, 14, 132, 1939.)

(micromeres) are formed at the animal pole. In the sea urchin eggs, the small cells formed during early cleavage are located at the vegetal pole. Instead of inventing a new name for these cells, the same descriptive term, "micromere", is used by embryologists, though it must be noted that it does not suggest homology (see Figure 6). Clearly, the size differences arise from peculiar positioning of the mitotic spindle. What cytoplasmic or surface properties determine the spindle orientation is unknown, though the pattern of cleavage in embryos is a very precisely controlled process that has been conserved in several major animal groups during evolution. One of the factors responsible for spindle orientation is some stratification of material in the egg cytoplasm with reference to the animal-vegetal axis.

However, the possibility of a role for the egg surface (especially the cortex) cannot be ruled out. Speksnijder et al.[20] have demonstrated the existence of an animal-vegetal polarity in the plasma membrane of the fertilized, uncleaved egg of the marine snail, *Nassarius reticulatus*. Using the technique of freeze-fracture, they have revealed the presence of intramembrane particles on the P and E faces of the plasma membrane. The main features of the plasma membrane from the different regions of the egg surface (Figure 7) are given in Table 1, which shows clearly a definite polarity. How the distribution of the microvilli or intramembrane particles is causally related to any developmental event is not clear. However, the existence of such a polarity is interesting. If similar regional differences are discovered in the eggs of different animal groups, we shall have the opportunity to compare them and see if any early developmental peculiarities are correlated with a particular pattern.

Recent studies on the early development of the mouse embryo have revealed some striking cell surface mechanisms that exert a decisive influence on the course of development. An important feature of the developing mammalian embryo is that the cleavage products are set apart, at an early stage, into two categories: those forming the inner cell mass and those developing into the trophoblast. The latter are specialized for the function of anchoring to the uterine tissue and invading it as the placenta develops. The inner cell mass develops into the embryo proper and, in addition, a number of extra-embryonic structures. As a consequence of the segregation, the inner cell mass is isolated effectively from the environment of the uterine lumen and tissues.

Following fertilization, as the mouse egg begins to divide, initially somewhat spherical

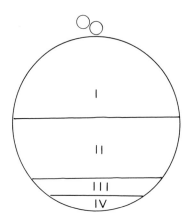

FIGURE 7.   The egg of *Nassarius* showing the demarkation of the surface zones studied by freeze fracture. (From Speksnijder, J. E., Mulder, M. M., Dohmen, M. R., Hage, W. J., Bluemink, J. G., *Dev. Biol.*, 108, 38, 1985. With permission.)

Table 1
MAIN FEATURES OF THE PLASMA MEMBRANE IN FOUR AREAS (I-IV)[a] OF THE *NASSARIUS* EGG AS VISUALIZED BY FREEZE-FRACTURE ELECTRON MICROSCOPY

Main features	Plasma membrane areas			
	I	**II**	**III**	**IV**
Microvilli pattern	Irregular	Regular	Rows	Irregular
Microvilli density	Moderate	High	High	Low
Total IMP[b] density, E[c]	Low	Moderate	Moderate	High
Total IMP density, P[c]	Moderate	Low	Moderate	High
E/P ratio	Moderate	High	Moderate	Moderate
Predominant IMP size	Large	Large	Small	Large
Significant difference in IMP size distribution with reference to other areas[d]	No	No	Yes	No

[a]   Egg surface areas as shown in Figure 7.
[b]   IMP, intramembrane particles.
[c]   E and P are the fracture membranes.
[d]   Tested by Kolmogorogov-Smirnov two-sample test.

From Speksnijder, J. E., Mulder, M. M., Dohmen, M. R., Hage, W. J., and Bluemink, J.G., *Dev. Biol.*, 108, 38, 1985. With permission.

cells are formed. When the embryo consists of eight cells, these cells show a marked change in their shape, forming a more "compact" embryo (Figure 8). It appears that the cells maximize their adhesion by assuming larger areas of contact surface. Since during this process the cells develop maximum contact, it is called compaction. Compaction has been shown to occur in rats, rabbits, monkeys, and humans also.[21] Determination of blastomeres to develop into the trophectoderm or the inner cell mass seems to occur after compaction. This is known from the experimental interchanging of the position of blastomeres at the eight-cell stage, which does not disturb normal development. Because compaction is a critical event in the development of the mouse blastocyst, it presumably is a general feature of all mammalian embryos.

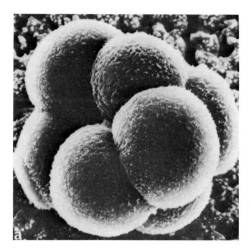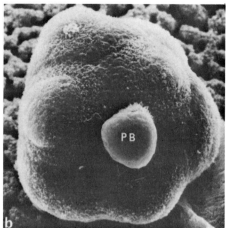

FIGURE 8. Scanning electron micrographs of uncompacted (a) and compacted (b) eight-cell mouse embryos. Note the change in cell shape and maximization of cell-cell contact. The polar body (PB) does not participate in compaction (b). (From Ducibella, T., *Development in Mammals*, Vol. 1, Johnson, M. M., Ed., North Holland, Amsterdam, 1977, 5. With permission.)

The cell surface changes associated with compaction have been receiving keen attention in the past 5 years or so. Compaction marks the onset of tight junction formation and thus provides for close cell-cell apposition and eventual development of occluding junctions at the morula stage (Figure 9). Maximization of contact between cells results in obscuring the cell junctions when observed with the light microscope. Reeve and Ziomek[22] have demonstrated that microvilli are distributed uniformly over the four-cell stage. At the eight-cell stage there is a progressive localization of the microvilli, so that they are restricted to the region normally exposed to the outside. Accompanying the shape changes are also cytoplasmic changes that reflect some polarization. Vital dye staining is intense between the nucleus and the exposed surface of the embryo.[23] Thus compaction seems to be accompanied by some polarization of the cells.

Wherever cell adhesion is involved, a role for Ca^{2+} can be assumed. Compaction of morula cells is no exception to this. It has been well established that calcium ions are necessary for compaction. Embryos in low Ca^{2+} media (0.02 mM) continue to undergo cleavage, but compaction does not occur. Even compacted embryos undergo "decompaction". Thus it appears that Ca^{2+} is essential for both compaction and maintenance of the compacted state. Experimentally compaction can be delayed until morula can recover (i.e., undergo compaction) when the normal Ca^{2+} level is restored.[21] It has also been shown that Ca^{2+}-dependent adhesion sites are involved in compaction. Besides the sites of Ca^{2+} action on the surface, there seems to be another. Recompaction of decompacted morulae is suppressed by inhibitors of Ca^{2+} uptake and calmodulin, suggesting an additional cytoplasmic site of action.[24] As time progresses after compaction, it becomes increasingly difficult to decompact embryos by exposing them to Ca^{2+}-free media. Presumably this is because the cell contacts are further stabilized by the development of junctional complexes.

The nature of the cell surface changes leading to compaction have been investigated using antibodies raised against a cell surface glycoprotein that occurs on embryocarcinoma cells.[25-28] Antibodies against embryocarcinoma cells were raised by immunizing rabbits. These antibodies and their Fab fragments prevent compaction of the 8 to 16 cell stage embryos and cause decompaction of such embryos that have already compacted. Thus the surface antigen on the embryonic cells seems to be causally related to compaction. The Fab fragment target on the embryocarcinoma cells has been purified and shown to be a glyco-

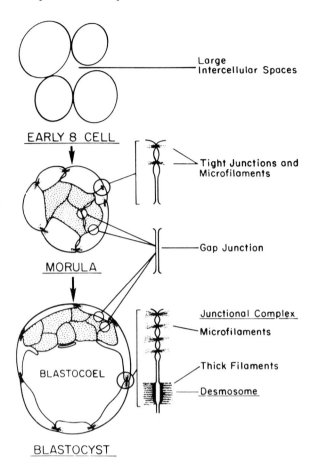

FIGURE 9.   Schematic summary of the development of intercellular junctions in the preimplantation mouse embryo. (From Ducibella, T., *Development in Mammals,* Vol. 1, Johnson, M. M., Ed., North Holland, Amsterdam, 1977, 5. With permission.)

protein of molecular weight ≈84,000 daltons.[27] A monoclonal antibody (ECCD-1), recognizing a cell surface component of ≈124,000 mol wt, has been shown to be able to decompact 8 to 16 cell stage mouse embryos. In these embryos, cell proliferation occurred normally and development of blastocyst-like vesicles was also attained. However, the embryo did not contain the inner cell mass. Shirayoshi et al.[29] have shown that the monoclonal antibody, ECCD-1, recognizes a Ca^{2+}-dependent cell adhesion mechanism which is essential for compaction.

Further work is needed to clarify which of the surface antigens are directly involved in adhesion and compaction of mouse morula cells. However it is now more or less certain that the cell surface mechanism leading to compaction depends on molecular recognition through carbohydrates. Earlier studies using lectin binding as a method of detecting surface carbohydrates had shown that polarization of cells during compaction is correlated with changes in the distribution of lectin binding sites.[26] Bird and Kimber[30] have recently investigated the question further. A mixture of lacto-*N*-fucopentaose II (in which fucose is linked α (1 to 4) to *N*-acetylglucosamine) and lacto-*N*-fucopentaose III (where the link is α [1 to 3]), can reverse the compaction of mouse morulae. However, other similar oligosaccharides that lack fucose or contain fucose linked α (1 to 2), or possess two fucose residues, do not cause decompaction. Sensitivity to lacto-*N*-fucopentaoses appears about the time that the embryos are becoming refractory to decompaction in Ca^{2+}-free medium. Thus the oligo-saccharides must be affecting the Ca^{2+}-independent phase of compaction. These studies provide strong evidence for the involvement of a saccharide-endogenous lectin interaction in compaction.

Beneath the closely apposed plasma membranes of compacting blastomeres, microtubules become localized and oriented parallel to the cell surface. Compaction depends on the action of microfilaments also as shown by the effect of cytochalasin B, which prevents compaction and decompacts the embryos already compacted.

Sequestering the inner cell mass suggests changes in the adhesiveness of morula cells.[31] Surani and Handyside[32] separated the inner and outer cells from 16-cell mouse morulae. Mixed cells from similar stages were also obtained. The cells were then fluorescently labeled. They were subsequently aggregated with unlabeled, partially decompacted 8 to 10 cell morulae. The position occupied by the labeled blastomeres within the aggregate was ex-amined after 3 to 6 hr in culture. It was revealed that most of the outer cells preferred an external position and flattened into extensive polygonal shapes. Unlike the outer cells, the inner cells preferred a deeper positioning. These results suggest that there are differences in the adhesiveness of the cells constituting the trophectoderm and the inner cell mass. Whether the difference in adhesiveness is the cause of their positioning cannot be considered proven. However the results of Surani and Handyside[32] suggest the possibility.

Differentiation of the inner cell mass seems to depend on its physical separation from the uterine environment. Similarly development of the trophoblast would be impossible if the blastomeres continued to retain their spherical shape. The early development of the mouse blastocyst described above involves highly organized cell surface changes in concert with cytoplasmic changes. These are geared to create conditions leading to the differentiation of the trophectoderm and the inner cell mass. Compaction of the blastomeres marks the onset of tight junction formation, consequently segregating the inner cell mass from the prospective trophectoderm. The "inner" environment seems to be imperative for the developmental divergence of the inner cell mass from the trophectoderm. The inner environment does not seem to be a mere chemical milieu; it seems to be cellular. Pedersen and Spindle[33] introduced early embryos (four cell or morula) into the blastocoel of a giant blastocyst obtained by fusing eight to ten uncompacted embryos. When a zona-intact embryo was introduced into the blastocoel, it developed into a normal blastocyst. Puncturing the zona did not alter the result, indicating that the chemical milieu of the blastocoel did not influence the development. However, when zona-free embryos were introduced, they developed into a cell cluster attached to the inner side of the host trophectoderm. The cell clusters had no blastocoel (Figure 10). This shows that contact with the inner aspect of the trophoblast cells has changed the developmental pattern of the embryo. Removal of such embryos, after 2 days stay in the blastocoel, allowed the growth of trophoblast, but no blastocoel was formed. These interesting experiments show that the cellular environment offered by the inner face of the trophectoderm causes the development of the inner cell mass.

Before we conclude the discussion on the role of the cell surface in the very early stage of development, viz., cleavage, another aspect of the problem may be highlighted. Descriptive studies on embryos have revealed a striking pattern of positioning of the cleavage products. The position of the cells formed during cleavage is not their definitive location. Is there then any significance to the relative positions of the cells of different prospective significance during cleavage? Do the cells communicate with each other, say, through exchange of chemical signals? Or, alternatively, do the cells remain together passively, without influencing each other? It is well-known in experimental embryology that removal of certain specific cells from the early embryos such as those of mollusks is reflected in a loss of some definite embryonic structures. However, the loss may be greater or less compared with the presumptive value of the cells ablated.

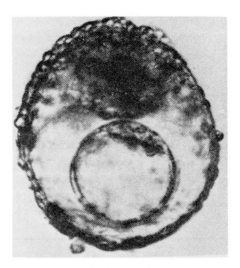

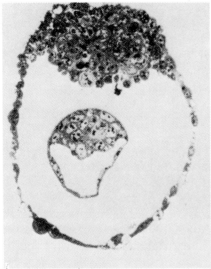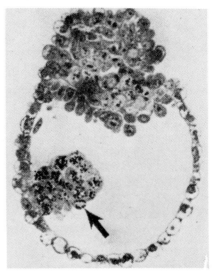

FIGURE 10.   The upper photograph shows a mouse blastocyst experimentally inserted into a giant blastocyst. *Lower left:* Section showing the normal development of a zona-intact blastocyst developing inside the giant blastocyst. *Lower right:* A zona-free blastocyst inserted similarly and shown in section. The inserted blastocyst was radioactively labeled and the cells derived from it (arrow) show silver grains in the autoradiograph. Note the difference between the organization of cells in the inserted blastocysts. (Reprinted by permission from *Nature (London)*, Vol. 284, p. 550. Copyright, 1980, Macmillan Journals Ltd.)

A general conclusion from these studies is that cells during early development, including the stages of cleavage, can communicate with each other.[34] Such communication is assumed to have developmental significance. The structural basis of the communication may be gap junctions. In fact, gap junctions are present between apposed cleavage cells of the mouse embryo as described in the previous paragraphs. The gap junctions provide a pathway of low electrical resistance. Transfer of small molecules across into a neighboring cell can (and does) occur through the gap junctions. Warner et al.[35] examined this question by a direct experimental approach. Gap junction components have been conservative during evolution, and hence antibodies raised against them from one animal can bind the antigens from another unrelated animal. Thus Warner et al.[35] have shown that antibodies obtained from rabbits

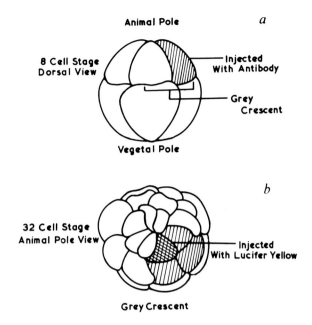

FIGURE 11. The location of the cell injected with antibody to gap-junctional proteins at the 8-cell stage (a) and subsequently with Lucifer yellow into a cell at the 32-cell stage (b). Hatched region indicates the presence of injected antibody. Cross-hatched region represents the presence of Lucifer yellow. (Reprinted by permission from *Nature,* Vol. 311, p. 127. Copyright, 1984, Macmillan Journals Ltd.)

immunized with rat liver gap junction proteins can bind similar antigens in the *Xenopus* egg and embryos. When the antibodies are injected intracellularly, they interfere with cell-cell communication, presumably by blocking the gap junctions by means of specific binding. The effective block of communication through the gap junctions following injection of the "antigap junction" antibodies is shown by the lack of electrical coupling between neighboring cells and the injected cells or their descendants. Besides, Lucifer yellow, a fluorescent dye which can diffuse across normal gap junctions, cannot be transferred from the antibody-injected cells into neighboring cells. Nonimmune rabbit serum does not influence intercellular communication, showing thereby that the effect of the antibodies is not nonspecific. Warner et al.[35] injected the antibodies into a particular identified cell (Figure 11) of the eight-cell stage *Xenopus* embryo. Communication between cells descended from this blastomere and their neighbors was blocked effectively. The injected embryos were allowed to grow further and examined for any malformations.

The most striking result of the study was that abnormalities were found in the regions related to the developmental fate of the injected cells. The most frequently occurring defects in such embryos were right/left asymmetries (Figure 12). The defects were not caused by cell death or other extraneous causes. It is clear therefore that the developmental defects had resulted from the failure of cell-cell communication through gap junctions in the cells affected. It is not known how long and through how many repeated cell divisions the injected antibodies remain effective. Nevertheless, these studies show conclusively that cellular communication through gap junctions is of developmental significance. There are many embryonic inductive stimuli exchanged between closely apposed cells. The exact nature of the signals exchanged is not known in any of the cases that have been studied so far. The findings of Warner et al.,[35] however, emphasize the need to reexamine these cases.

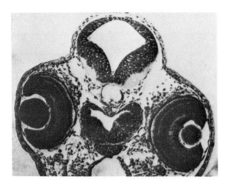 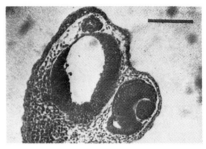

FIGURE 12. Transverse sections at the level of the developing eyes in a control embryo (left) and an embryo of the same age that had been injected with the antibody to gap-junctional proteins as in Figure 11 (a). Scale bar, 500 μm. The experimental embryo shows three features of note: (1) the eye is missing on the side previously injected with the antibody; (2) the brain is the shape normally found anterior to the eyes, and (3) the brain is particularly underdeveloped on the injected side. (Reprinted by permission from *Nature (London)*, Vol. 311, p. 127. Copyright 1984, Macmillan Journals Ltd.)

III. GASTRULATION

Repeated divisions of the fertilized egg give rise to a multicellular structure known as blastula. In those eggs which do not divide completely, a cellular blastodisc is formed. In any event, the cells thus formed are not in their definitive locations. A large scale rearrangement of the cells then occurs, leading to the establishment of germ layers. It occurs as a critical developmental event in the embryos of all multicellular animals. On this score, it is a developmental change of fundamental importance. Insofar as the outcome of this cellular rearrangement is the same, viz., the definitive positioning of the germ layers, one may expect to find highly conserved cellular mechanisms of achieving it in all multicellular animals. However, this is not so. In spite of the striking unity of plan in the construction of the chordate body, the pattern of cellular rearrangement during gastrulation varies widely in different chordates. This is partly due to differences in the yolk content of the eggs. However there seem to be other factors which are yet to be elucidated. Eggs that have very little yolk generally have a hollow blastula. In such embryos, one side of the hollow ball flattens and eventually the flat region sinks in without breaking at the rim. This is the simple process of invagination. In the eggs that have a moderate amount of yolk, the cleavage products are widely varying in size, those at the vegetal pole being the largest. In such embryos, the larger cells are not active in invagination; instead, the smaller cells spread over the larger ones. Finally, in the eggs in which a blastodisc is formed, the process of cellular rearrangement is further modified. We shall elaborate on this in a later section.

Many distinct cellular activities are involved in the restructuring of a blastula into the next stage called gastrula. The entire process of transformation of a blastula into a gastrula is called gastrulation. However, embryologists studying the process in different types of embryos often use the term in a somewhat restricted sense, referring only to the most striking phase in it. A number of terms are in current use to describe the collective activities of the cells during gastrulation. They are (1) ingression, (2) delamination, (3) invagination, (4) involution, and (5) epiboly. Briefly, these terms are explained below:

Ingression — This is a pattern of cellular rearrangement whereby individual cells leave the wall of the blastocoel and are segregated as single cells or as a separate mass. It may occur in a single location or several locations, thus distinguishing between unipolar and multipolar ingression.

Delamination — The separation of a group of cells as a distinct layer is known as delamination.

Invagination — When the wall of a hollow vesicle is folded in so as to form a double layered cup, the process is called invagination.

Involution — When the edge of a sheet of cells moves back on itself, turning over at a certain point, the process is termed involution. It must be noted that invagination is always accompanied by involution to a greater or lesser degree. In fact, it is difficult to cite examples of gastrulation wherein invagination or involution occur exclusively.

Epiboly — This is the process in which a group of cells spreads over another.

These terms are sometimes used with reference to developmental processes other than gastrulation also.

A. Sea Urchins

The sea urchin eggs are small and spherical with very little yolk content. Cleavage is holoblastic. The first five cleavages are illustrated in Figure 6. The first cleavage passes from the animal to vegetal pole, forming equal sized blastomeres. During the second cleavage, which occurs at right angles to the first, four cells of equal size are formed. In the third cleavage, which is horizontal, eight blastomeres of equal size are formed: four of them are the animal blastomeres and the other four vegetal blastomeres, situated towards the animal and vegetal pole, respectively. The next cleavage is vertical in the animal half and horizontal in the vegetal. The eight blastomeres of the animal half are of equal size. However, those at the vegetal half are of very unequal size, as shown in Figure 6. The next cleavage is vertical in the vegetal half and horizontal in the animal half. Subsequent cleavages are more and more asynchronous. A blastocoel develops and the embryo is now designated a blastula. It is a hollow sphere with its wall made of a single cell layer. Cilia develop on the surface of the blastula. After hatching from the egg membrane, it swims in the sea water. Gastrulation may be considered to have commenced when the cells derived from the micromeres begin to move towards the blastocoel. The cells thus immigrating by a process of ingression constitute the primary mesenchyme. Apparently these cells become less adhesive and begin to move into the blastocoel. Certainly there must be something to ensure that they move in rather than out.

In a recent study, Fink and McClay[37] have analyzed the changing affinities of the 16-cell stage micromeres and their descendants, which were grown in culture, and also of the primary mesenchyme cells, which have ingressed. The affinity of these cells was tested against three substrata: (1) hyalin, which is the major protein component of the extracellular hyaline layer on the outer surface of the blastula cells; (2) gastrula cells (other than primary mesenchyme) in monolayer; and (3) the basement membrane. The affinity of micromeres and their descendants for hyalin adsorbed onto microtiter test plates revealed an interesting change during development. Initially the micromeres exhibited affinity for the adsorbed hyalin, but then lost it at a developmental time corresponding to the beginning of ingression. Affinity of the micromeres for a monolayer of gastrula cells also showed a similar change, decreasing as the development progresses. On the other hand, the micromeres showed an increased affinity for the basement membrane with developmental time. Presumptive ectodermal and endodermal cells did not show such affinity. These observations suggest a simple cellular basis for the ingression of the primary mesenchyme during sea urchin gastrulation. The changing affinity could itself account for the cells' movement in rather than out.

Specialization of the surface of micromeres seems to set in very early in development. De Simone and Spiegel[38] have reported that cell type specificity of surface proteins sets in from the 16-cell stage of the sea urchin embryo. Four unique cell surface proteins of the micromeres were identified in an SDS-PAGE analysis of proteins radioiodinated in intact

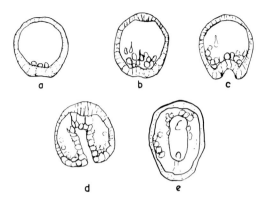

FIGURE 13.   Diagrams (a to e) showing the process of gastrulation in the sea urchin embryos. Note the ingression of the primary mesenchyme cells (a) and (b) and invagination of the cells lining the archenteron (c to e).

blastulae of *Arbacia punctulata*. In *Strongylocentrotus droebachiensis*, a micromere-specific protein of molecular weight ≈133,000 was identified. This protein was found to bind WGA but not Con A. This observation clearly implicates the surface protein in cell recognition processes. It appears that micromeres begin to differentiate at the 16-cell stage, and their differentiation reaches the ''right'' stage when ingression is to begin.

In the blastocoel, the primary mesenchyme cells move as free cells with long, fine, and branched filopodia, which attach at the junctions between ectodermal cells. It is obvious that the migrating primary mesenchyme cells use the inner wall of the blastula for anchorage. However, there is some sulfated glycosaminoglycan material in the blastocoel fluid that may exert definite influence on the migrating cells.[39]

More or less concurrently with the ingression of the primary mesenchyme cells, the vegetal region of the blastula flattens and then begins to invaginate. An archenteron is thus formed (Figure 13). This process of invagination continues until the archenteron extends to about one third of the blastocoel cavity. What initiates the invagination is still obscure. One can think of a mechanism in which the inner ends of the presumptive endodermal cells become less adhesive so that the early invagination is autonomous, i.e., it is not controlled by other parts of the embryo. This accords with the changing affinities of the primary mesenchymal cells as described by Fink and McClay.[37] When the animal half of the blastula is separated from the vegetal half, invagination begins, nevertheless. Whatever the factors for the initiation of invagination may be, they are confined to the vegetal plate. After a short pause, invagination continues further. Cells at the tip of the archenteron then protrude long filopodia, which adhere to the inner side of the apical pole (animal pole) and the future dorsal side. The filopodia seem to be contractile, and by exerting tension, they appear to pull the tip of the archenteron to the animal pole (Figure 13). The attachment of filopodia to the animal pole and their contraction seem to be causally related to the completion of invagination after it is stopped at about one third of the blastocoel.[40,41] It cannot be concluded, however, that invagination in all animal embryos is brought about by anchoring filopodia. A voluminous fibrillar extracellular matrix composed of filaments, twisting fibers, and granules lining the blastocoel of midgastrula embryos has been described.[42] The matrix material is closely associated with the basement membrane of the ectodermal cells. Histochemical studies have revealed that the matrix is composed mostly of sulfated glycosaminoglycans. It has been suggested that the fibrillar network is also responsible for guiding the invaginating endodermal cells.

B. Amphibia

The amphibian egg is characterized by a moderate yolk content. Cleavage results in larger vegetal blastomeres and smaller animal blastomeres. The size differences presumably arise from a gradient of yolk content in the egg, decreasing in the direction of the animal pole. Mitosis in the yolky cells is tardy. As a consequence, the vegetal blastomeres remain larger. Whether the tardy cleavage in the vegetal region is solely due to the yolk content is not clear. An additional feature of the amphibian blastula is that the blastocoel wall is several cell layers thick. An active participation of the presumptive endodermal cells (approximately the whole of vegetal half) in gastrulation is minimized, presumably because of their bulkiness. Instead, the smaller cells of the animal pole spread over the larger cells (epiboly). In the beginning of gastrulation, if the embryo is observed from the surface, the animal half of it looks darker because of the presence of pigmented cells. The vegetal half, consisting of the yolk-laden endodermal cells, appears paler. An intermediate zone known as the marginal zone is also recognized. It extends over the equator, going all around the spherical embryo. As the embryonic axes are already determined, the terms dorsal, ventral, and lateral marginal zone can be used for more precise reference. As gastrulation proceeds, the marginal zone spreads over the vegetal half. The crescent shaped dorsal blastoporal lip, which marks the beginning of gastrulation, extends laterally as the process of epiboly continues. Finally, the blastoporal lip becomes circular, enclosing a circular area of the vegetal cells (yolk plug).

Amphibian gastrulation combines epiboly with invagination and involution. The existence of the last two processes was demonstrated by the classical experiments of Vogt.[43,44] Vital dye marks were made on the surface of an early gastrula to stain the superficial cells. The dye does not diffuse from cell to cell and hence the marked cells could be followed without ambiguity during their movement. The experiments of Vogt have been repeated by many embryologists and generally confirmed in a variety of amphibian species. If a dye mark is placed on the middorsal line (see Figure 14) just anterior to the dorsal blastoporal lip, it appears to stream towards the crescent shaped invagination center. The two lateral spots (2 and 2′) also follow a similar course, though more slowly and, in addition, tending to converge towards the invagination center that in the meanwhile has extended laterally. Eventually the dye marks assume an internal position showing thereby that the marked cells have turned in and formed a new layer. Detailed studies on these lines have revealed the trajectories of cell movements during amphibian gastrulation (Figure 14). Although there are some matters of finer detail about which there is some uncertainty, the essential features of gastrulation elucidated by the dye marking experiments are correct.

Based on the vital dye marking experiments and histological studies, the collective activity of cells during epiboly has been described as spreading of the presumptive ectodermal cells over the presumptive endodermal cells. The deformations of individual cells are obviously integrated in a complex manner to bring about the process.

Recent studies on *Xenopus* embryos have added some more detailed information on the cellular activities during gastrulation. Keller[45] has described the process of epiboly and invagination using scanning electron microscopy of fixed, fractured embryos. Changing cell dimensions during the progress of gastrulation have also been analyzed in an attempt to offer a mechanical explanation of gastrulation. The animal half, consisting of the presumptive ectoderm of the embryo, is a multilayered cellular structure. The outermost layer of cells is epithelial and constitutes the "superficial cells". The deeper nonepithelial cells are distinguished into two types. The "inner deep cells" are those that bound the blastocoel or the involuted mesoderm. The "inter deep cells" are those deep cells that are between the superficial cells and the inner deep cells. During epiboly, the cells change their shape in a characteristic manner. The superficial cells flatten. The inter and inner deep cells interdigitate (Figure 15) as they also change from a more columnar shape to cuboidal. When interdigitation is complete, the deep cells are one layer thick. The result of these changes is spreading of

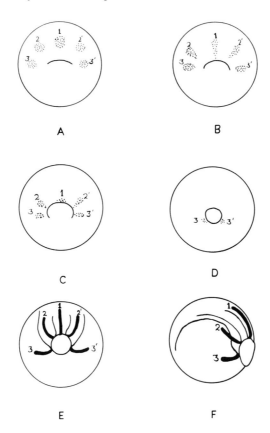

FIGURE 14.   Diagram illustrating the vital dye marking experiments of
Vogt.[43,44] The vital dye marks (1, 2, 2', 3, and 3') made on the surface
of the amphibian egg, as shown in (A) change their shape and location
with reference to the blastopore which is also changing concomitantly from
a crescent shape to a circle (A to D); (E) and (F) trajectories of the
displacement of the different vital dye mark shown in posterior view (E)
and profile (F). The thick lines represent the surface, and the thin lines
the interior displacements.

the multilayered animal half region over the presumptive endoderm. In the region of the
dorsal marginal zone, these changes occur rapidly. Similar changes occur in the other regions,
but to a lesser extent, and start late in developmental time. Increase in the area and spreading
of the prospective ectodermal cells is largely due to interdigitation of the cells; flattening of
cells contributes much less. Further, there seems to be very little mitotic activity in the
region. In any event, cell division does not act as a primary morphogenetic force during
epiboly.[46]

The process of invagination during amphibian gastrulation has been a favorite of many
embryologists during the past several decades. As a result, detailed information on the
changes occurring during the gastrulation of many amphibian species is now available. The
first indication of gastrulation in the frog or newt is the appearance of a faint, crescent
shaped groove or depression at a point that generally coincides with the dorsal side of the
embryo and the meridian marked by the grey crescent. The cells here seem to stream in,
developing long narrow necks at the outer (apical) ends and rotund bases towards the
blastocoel. These cells are called "bottle cells" or "flask cells" because of their shape
(Figure 16). Rhumbler,[47] who first described the characteristic shape of these cells, concluded
that they initiate the process of invagination. Subsequent workers found some evidence to

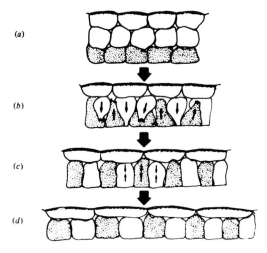

FIGURE 15.   Diagrams to show the mechanism of spreading of the deep region of the *Xenopus* gastrula during epiboly and extension. From the outside inward, three layers of cells are shown: (a), the superficial layer, the interdeep layer, and the inner deep layer (shaded). Cells of the two deep layers extend protrusions inward or outward, along radii of the embryo, and move between one another (arrows, b); this process is called radial interdigitation. When interdigitation is complete, the inner deep layer consists of one layer of elongated, columnar deep cells (c). These cells then flatten and spread (d). Cell division occurs as well, but has not been shown for clarity of illustration. (From Keller, R. E., The cellular basis of epiboly: an SEM study of deep cell rearrangement during gastrulation in *Xenopus laevis, J. Embryol. Exp. Morphol.*, 60, 201, 1980. With permission of the publisher, the Company of Biologists, Cambridge University Press.)

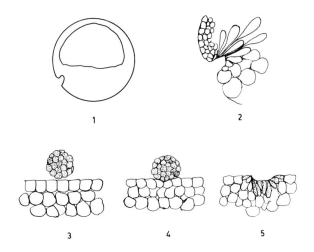

FIGURE 16.   Bottle-shaped cells at the amphibian blastoporal lip: (1) diagram of a vertical section of early gastrula showing the indentation on the left, marking the invagination center; (2) diagram depicting the shapes of cells at the invagination center; (3 to 5) Holtfreter's experiment,[49] showing the changing shape of amphibian blastoporal cells (shaded) into an endodermal substratum.

support this conclusion. Holtfreter[48,49] demonstrated that when the surface ectodermal cells of the blastopore region are placed on a mass of internal (blastocoelic) endoderm cells in vitro, the former exhibit a morphogenetic activity remarkably similar to that observed in vivo. The surface cells assume the characteristic bottle shape and "invaginate" into the cellular mass consisting of the deeper endodermal cells (Figure 16). This observation is

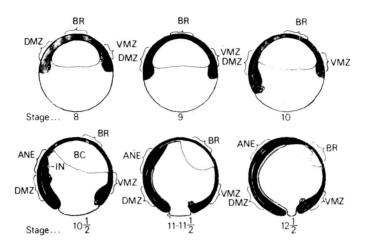

FIGURE 17.   Diagram representing changes in the maginal zones during gastrulation in *Xenopus*. Two components can be distinguished in the gastrula — an outer shell (lightly shaded), which undergoes spreading or expansion toward the blastopore, and an inner region (darkly shaded), which has undergone involution and is clearly separated from the outer shell by an interface (IN). The diagrams show the changes in the morphology and arrangement of cells in the following sectors of the outer shell of the embryo: BR, blastocoel roof; DMZ, dorsal marginal zone; VMZ, ventral marginal zone, and ANE, anterior neural ectoderm. (From Keller, R. E., The cellular basis of epiboly: an SEM study of deep cell rearrangement during gastrulation in *Xenopus laevis, J. Embryol. Exp. Morphol.*, 60, 201, 1980. With permission of the publisher, the Company of Biologists, Cambridge University Press.)

generally considered to suggest that the bottle cells, by assuming their peculiar shape, generate a stress on the superficial layer of cells and thereby cause invagination. Whether they have a principal role in amphibian invagination is, however, subject to debate.

Keller[50] examined the question in *Xenopus* embryos. If the bottle cells are removed partly or even completely, involution still occurs. In other words, the head endoderm and the underlying presumptive mesoderm in the deep dorsal marginal zone (see Figure 17) move in, and they assume their definitive position even in the absence of the bottle cells. If the deep region of the dorsal marginal zone is disturbed, involution of the deep cells, as well as superficial head endodermal cells, is inhibited. If the deep marginal zone is replaced with a patch of the deep region from the blastocoel roof, no involution occurs. Keller's[50] experiments thus show that the cells of the deep region of the marginal zone have unique properties, which are necessary for involution. The role of the bottle cells is, according to this interpretation, restricted to the initiation of the process and does not aid its continuation. According to these results, amphibian gastrulation involves invagination only to a very limited extent.

Before one can visualize a cellular basis of amphibian gastrulation, several questions have to be answered. Are gastrulation movements controlled by the embryo as a whole or by the individual cells? Are the cells in various zones of the blastula already programed to show the changes according to a definite time course irrespective of what happens to the other neighboring zones? It is not easy to get categorical answers to such questions at present. However, there are indications that a certain degree of autonomy is possessed by the cells. Excised deep cells of the marginal zone, prior to involution, actively interdigitate between cells of the inner surface of blastocoel or the marginal zone.[45] Le Blanc and Brick[51] analyzed the embryonic cells by scanning electron microscopy and revealed some interesting differences among the various cell types. The presumptive epidermal cells were found to spread and flatten, thus showing a tendency similar to that in epiboly. The invaginating cells of the presumptive head endoderm of the early gastrula elongate and produce filopodia somewhat reminiscent of their deformations in vivo. Cells of the different regions also show differences in electrophoretic mobilities.[52]

Interesting as these observations are, they do not explain the cellular mechanisms involved in gastrulation. What still remains almost mysterious is the beautiful orchestration of the whole process. Cells in the different regions of the blastula seem to have a detailed plan defining how far to go and when. Once the process starts, everything usually goes according to the plan, each step of which is meticulously coordinated. Even if the vitalistic overtones of these statements are discarded, what remains of them is still true. In order to find a more rational explanation, one ought to identify the discrete cellular and extracellular mechanisms. An important role is obviously played by the cytoskeletal structures, especially the aspects of their assembly and disassembly. The cell surface is undoubtedly going to be the most important object of future studies aimed at revealing the mechanisms. It is already known that some recognition system involving cell surface carbohydrates is one of the components of the mechanism.[53] The spatio-temporal nature of the mechanism can be understood on the basis of gradients of distribution of substances. But what these substances are and what pattern of distribution exists is unknown.

Directional movement of cells such as that involved in the migration of presumptive mesodermal cells during involution can be controlled by materials in the extracellular matrix. A network of extracellular fibrils on the inside of the ectodermal layer has been demonstrated.[54] The network can also be transferred on to the surface of a coverslip. The surface of the coverslip thus conditioned can promote adhesion and locomotion of prospective mesodermal cells of the blastula.[55] In vivo, however, the migrating mesodermal cells show a characteristic orientation towards the animal pole, as revealed by scanning electron microscopy. A casual relation between the fibrils and the vectorial migration of cells is not evident, though it probably exists. It has been shown that the fibrillar network consists of fibronectin, which can be revealed by an immunolabeling method specific for the matrix component.[56,57] No obvious orientation is observed in this network, which is mostly composed of a three-dimensional array of fibers. Several experimental observations suggest a definite role for the fibronectin network in amphibian gastrulation. It is absent in interspecific hybrid embryos that are arrested at gastrulation. A correlation between the absence of the fibers and failure of gastrulation is thus clear.[55] If the ectodermal piece on which the involuting mesodermal cells migrate is excised and replaced after rotating so as to offer its outer surface for migration, no mesodermal cells adhere to it.[57] Specially striking is the observation that when antifibronectin monovalent antibodies (Fab) are injected into the blastocoel of early blastula, no gastrulation occurs. Preimmune Fab or antifibronectin Fab neutralized with fibronectin has no such effect, showing clearly that the effect is specifically due to fibronectin (Figure 18).[56] Further, a competitive peptide inhibitor of fibronectin inhibits gastrulation.[58]

It is clear from the recent findings discussed above that the mechanism of involution during amphibian gastrulation is beginning to be understood. An important component of this is a fibronectin network. Another important component of the mechanism is, however, still missing. The migration of the presumptive mesodermal cells is highly patterned, and there must be something controlling it very precisely. Probably the fibronectin network is working in concert with some other component(s) of the extracellular matrix. Nakatsuji and Johnson[55] have reported that coverslips can be conditioned by pieces of explanted blastocoel roof ectoderm held during culture in such a manner that they are under tension in a certain direction. Cells moving on such coverslips do so in an oriented manner corresponding to the direction of tension in which the ectodermal sheet was held during conditioning. A good correlation between the two parameters was found by image analysis of the cell trails with a computer. No oriented movement of mesodermal cells was observed if the ectodermal sheet was held without tension in any direction during conditioning of the coverslip. There is, however, no evidence to show that the fibronectin has anything to do with the orientation of cell movement. It could as well be that the ectodermal sheet lays down some other component of the extracellular matrix, which may be responsible for the contact guidance

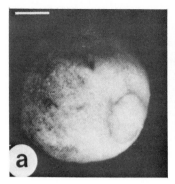
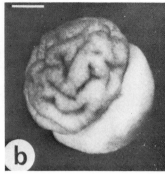

FIGURE 18.   Intrablastocoelic injection of monovalent antibodies to fibronectin at the blastula stage of *Pleurodeles Waltlii:* (a) control embryo injected with antibodies to fibronectin preincubated with the antigen, develops normally; (b) experimental embryo observed 24 hr after injection with antibodies to fibronectin. Gastrulation is completely inhibited and the blastocoel roof becomes disorganized. Bar, 0.3 mm. (From Thiery, J.-P., Duband, J.-L., Tucker, G., Darbiere, T., and Boucaut, J.-C., *Prog. Clin. Biol. Res.,* 151, 187, 1984. With permission.)

phenomenon observed by Nakatsuji et al.[59] One thing is, however, clear: mere physical forces such as tension on the cells may be the primary cause of a certain spatial pattern exhibited in cell behavior. The striking pattern of cell movement during gastrulation may be primarily based on such physical forces, which have to be elucidated by future work.

C. The Chick Embryo

An early event in the segregation of cleavage cells into distinct germ layers in the avian embryo is the formation of the primary hypoblast. It is formed by a gradual separation of individual cells from the lower strata of the blastoderm. This process occurs more predominantly in the future posterior part of the blastoderm. Initially the cells constituting the primary hypoblast are somewhat scattered, i.e., the layer formed by them is discontinuous, particularly so at the anterior end. Eventually, however, a continuous layer is formed. It may be noted that this is a typical process of ingression. The cells of the primary hypoblast have been shown to migrate radially.[62] This suggests that the under surface of the epiblast provides some structural pattern that dictates the movement. Subsequently, when the primitive streak is formed, cells proliferating from its tip begin to invade the primary hypoblast in the region of the Hensen's node. Gradually the cells constituting the primary hypoblast are displaced towards the periphery, their place being taken by the ingressing cells from the tip of the primitive streak. The primary hypoblast gives rise to the extra-embryonic endoderm of the yolk sac whereas the newly formed lower layer, viz., the definitive endoblast, is the source of the embryonic endoderm and part of the extra-embryonic endoderm.

The evidence leading to the above conclusion was gathered gradually over the last 25 years. Early embryologists who studied the process using vital dye marks[60,61] could not distinguish between the primary hypoblast and definitive endoblast. Later Spratt and Haas[62] demonstrated extensive and highly organized movements of cells in the lower layer of the chick blastoderm using carbon and carmine particles for tracking the moving cells. These movements are radial, i.e., from the center to the periphery. Autoradiographic methods[63-66] clearly showed that the hypoblast of the early chick blastoderm is replaced by the definitive endoblast, which therefore is to be regarded as a distinct layer with a different origin and developmental fate. Finally, the use of the "biological marker" of quail heterochromatin helped further elucidation of the process of endoderm formation in the avian embryos.[67]

Table 2
PROPERTIES OF CHICK EMBRYONIC PRIMARY HYPOBLAST AND DEFINITIVE ENDOBLAST CELLS

Primary hypoblast	Definitive endoblast
1. Rounded and loosely held together; presumably have low mutual adhesive strength; the restricted regions of mutual contact have no specialized membrane junctions	Closely packed; presumably have strong mutual adhesion; tight junctions between cells
2. In explants, cells move rapidly ($\sim$1.8 μm/min); have bigger regions of ruffling	In explants, cells move slowly; limited ruffling activity
3. Contact inhibition is not pronounced	Pronounced contact inhibition of locomotion
4. Cells detach from explants	Cells do not detach from explants
5. Cells at the edge of explants intermingle with cells of definitive endoblast in confrontation cultures	Cells at the edge of explants do not intermingle with homotypic cells in confrontation cultures

Based on Sanders, E. J., Bellairs, R., and Portch, P. A., *J. Embryol. Exp. Morphol.*, 46, 187, 1978.

The cellular activities in the formation of the primary hypoblast and the definitive endoblast have been investigated recently. An important aspect of the process of hypoblast formation is that individual cells located in and around the Hensen's node presumably lose mutual adhesiveness and somehow acquire limited motility so that they can "drop down" and form a lower layer. It is not known if they change their affinity for neighboring cells and for any extracellular material in a differential manner so as to orientate their "dropping down" and not moving up. The other aspect is related to the radial displacement of the primary hypoblast by the definitive endoblast. The cellular properties involved in this process have been analyzed by Sanders et al.[68] and Bellairs et al.[69] It has been shown that the primary hypoblast and definitive endoblast cells differ in some important ways which seem to be relevant to their behavior in morphogenesis (Table 2). The definitive endoblast cells seem to be more strongly adhesive and capable of penetrating into a sheet of primary hypoblast cells which consequently assume a peripheral position. It may be recalled here from the previous chapter that in experimental models of cell sorting, the more adhesive cells are assumed to displace the cells of weaker adhesiveness to a peripheral position. It is quite possible that the hypoblast of the avian embryo offers a relatively loosely organized epithelium with small transient gaps which may be made use of by the ingressing cells.[69]

The next important aspect of avian gastrulation is the formation of the middle layer or mesoderm, which is inserted between the upper and lower layers. During this process, the cells of the epiblast migrate medially and through the primitive streak and then again laterally, thus forming the middle layer (Figure 19). It is not a process of typical invagination or involution, since the cells do not migrate forming a cavity resembling the amphibian archenteron. One of the older controversies regarding avian gastrulation was whether the presumptive mesoderm cells take a surface-to-deep layer course or do they just proliferate from the primitive streak that may act as a "growth center".[70] It is now clear that the controversy was generated by the observation that carbon or carmine particles, used as cell markers and placed on the upper surface of the chick blastoderm, do not pass through the primitive streak and assume a deeper position. In the first place, large, microscopically visible particles are not likely to be carried into regions where close cell contacts exist. Thus the marker particles would not pass from the surface to a deeper layer even if the cells do. Besides, there is now convincing evidence[71] that the presumptive mesoderm cells do undergo surface-to-deep layer migration.

Important changes in the cell surface and intercellular matrix must be occurring at the primitive streak where the cells change their morphology from an epithelial to a mesenchymal

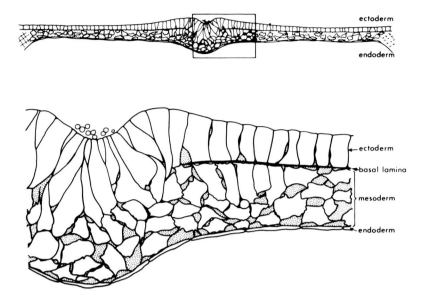

FIGURE 19.   Drawings to show the shapes of cells in the chick blastoderm at the primitive streak. The boxed part of the upper figure is shown enlarged below. (Courtesy of Professor Ruth Bellairs.)

form. Sanders,[72] in a recent investigation, has reported that the basement membrane of the early chick blastoderm can be labeled locally in a small region by injecting ferritin-conjugated Con A, which can then be detected by transmission electron microscopy. The label does not diffuse to other cells, and hence interpretation of cell movements based on the displacement of ferritin-Con A labeled cells is reliable. Sanders[72] has shown that the medially migrating epiblast cells move along with their labeled basement membranes. Thus the medial migration of the epiblast occurs as a sheet of cells. At the primitive streak, the basement membrane breaks down (Figure 19). Disruption of fibronectin and laminin organization of the basement membrane of the epiblast cells occurs as they migrate through the primitive streak.[73,74] An interesting question for further investigation is regarding the mechanism and control of the breakdown of the basement membrane. After all, a highly organized structure such as the basement membrane cannot break down without some specific mechanism. Besides, the breakdown is a highly controlled and localized event, restricted only to the region of the primitive streak.

During their lateral movement, some mesodermal cells seem to use the basement membrane of the overlying epiblast as a substratum. As in the amphibians, fibronectin and other components of the extracellular matrix have been implicated in the control of the process. By immunolocalization, fibronectin has been demonstrated in the basement membrane of the upper layer.[75] At the primitive streak where mesodermal cells migrate inward, fibronectin is absent. Similarly, the migrating mesodermal cells are devoid of fibronectin.

Harrisson et al.[76] employed two different techniques for the detection of fibronectin in the early chick embryo: (1) ethanol/acetic acid fixation, or (2) formaldehyde/glutaraldehyde fixation, followed by immunocytochemical detection in sections by means of the peroxidase/antiperoxidase method of Sternberger et al.[77] Aldehyde fixation demonstrated the ubiquitous presence of fibronectin in the whole basement membrane. In contrast, ethanol-fixed material led to masking of antigenic sites of fibronectin at the epiblast-mesenchymal interface. The antigenic sites were partly exposed by a treatment with hyaluronidase. According to these observations, there is a definite regional pattern of reaction between fibronectin and glycosaminoglycans. Harrisson et al.[76] suggest that there is an interaction between the two components of the extracellular matrix at that particular part of the basement membrane where the mesodermal cells have migrated. Whether the epiblast modifies the basement membrane allowing the mesodermal cells to migrate over it or the mesodermal cells induce the changes is not known. It is, however, interesting to note that some interaction between the components of the matrix is correlated with the migration of cells. In this context, we may recall what was said regarding the orchestration of events during amphibian gastrulation. The work of Harrisson et al.[76] clearly indicates how such a precisely timed and spatially organized process as gastrulation can be controlled by the different components of the extracellular matrix through their interactions among themselves and with the cells.

IV. IMPLANTATION

A unique feature of the development of eutherian mammals is viviparity, which involves the retention of the fertilized egg in the maternal uterus until the embryo has developed. Though viviparity has been developed in many groups of invertebrates and in all classes of vertebrates except the birds,[78] it has received most attention in eutherian mammals, for obvious reasons. Physiologically it is a very intricate developmental adaptation in which a number of hormones act in unison. Any disturbance in the delicate hormonal control of the process is likely to result in abortion. Most of the research in this area has been pursued by endocrinologists. It is not practicable here to give even a brief summary of the knowledge gained in this vast field. The following account is intended merely to highlight some of the developmental events wherein changing cell surface properties may be implicated. It is necessary to emphasize here that both the maternal and fetal tissues undergo precisely programed changes that are geared to bring about a successful pregnancy and to terminate it at the right time.

Earlier in this chapter we have referred to the early development of the mammalian egg. When the cleavage cells are segregated into the trophectoderm and inner cell mass, a fundamental dichotomy is established between the developing embryo and the maternal organism on the one hand, and those destined to form all the embryonic and most of the extraembryonic structures on the other. Up to this stage, the mammalian embryo is covered by the zona pellucida of the egg. For implantation, i.e., the histologically intimate apposition of the conceptus to the uterine wall, the blastocyst has to be released from the zona. Shedding of the zona seems to be controlled, at least partly, by the secretions of the uterus.[79] Some of the early events of blastocyst development, including its release from the zona, can be studied in vitro, in a culture medium supplemented with fetal calf serum. In simpler media lacking amino acids and serum, rupture of the zona is delayed. The delay can be eliminated when such simple media are supplemented with amino acids, vitamins, and fetuin, which is an α-globulin extracted from the serum.[79] How "nutrients" could break down the zona is, however, not easy to understand. Rosenfield and Joshi[80] have demonstrated endopeptidase activity in the rat uterine fluid having a lytic effect on the zona. This may well be a general feature of all eutherian mammals. Clearly, in vivo the uterine fluid is the source of the factor(s) required for the release of the blastocyst.

The next critical event in the development of a blastocyst is its specific adhesion to the uterine tissue. It is surprising that a developing blastocyst can adhere to tissues in many ectopic sites where implantation under any circumstance cannot be imagined to occur. Thus a blastocyst can show typical changes of early implantation in the anterior chamber of the eye. In fact the uterus is the least receptive tissue to a blastocyst, except during the normal phase of reproductive cycle when implantation is scheduled to occur. During this phase, a complex hormonal mechanism comes into action leading to an elaborate preparation of the uterus for implantation. The initial attachment of the blastocyst on the uterine epithelium is facilitated by changes occurring in the uterus as well as the trophectoderm. Unlike ectopic attachment of the trophectoderm, its adhesion to the uterine wall depends on mutual surface interactions. It seems that specific changes occur on both the embryonic and maternal cell surfaces to overcome the mutual nonadhesiveness, which is a typical attribute of the apical ends of epithelial cells.[81] Attachment and spreading behavior of trophectoderm to polystyrine or other inert surfaces has been used in many studies as an experimental model. Yet it seems that the experimental model does not adequately duplicate the in vivo conditions.

It may be considered as proven that the trophectoderm has the cells specialized for implantation, and the inner cell mass has no role in the process, at least in the initial stage. Isolated trophectoderm or small vesicles of it can implant, but not isolated inner cell masses. Empirically, two phases of the initial blastocyst-uterine contact have been recognized by some workers:[82,83] apposition and adhesion. The change seems to occur between day 5 and 6 after fertilization in the rat embryos. During the first phase, viz., apposition, the blastocyst seems to be a passive participant. Some changes take place during the development from day 5 to 6. Presumably these changes are functionally important. Ultrastructural studies have shown that during apposition, the trophectoderm cells have microvilli that interdigitate with those on the uterine epithelial cells. At the time of adhesion, the cell surfaces are flattened and parallel, tending to resist separation.[83] A decrease in the surface negative charge, inferred from decreased binding of cationized ferritin, coincides with the adhesion of trophectoderm.[84] Developmentally regulated changes in the surface charge of the blastocyst and uterine cells have been reported by other workers also.[85] Though a clearly defined causal relation between the changing surface charge and implantation events has not yet emerged, such a possibility seems attractive. Lectin binding properties of the blastocyst surface also change, coinciding with the transition from apposition to adhesion (Table 3).[83] These facts indicate that the blastocyst adheres to the uterine surface through a mechanism of cell recognition involving cell surface carbohydrates.

In spite of the fact that the trophectoderm can show the early changes of implantation in a variety of experimental conditions in the absence of a uterine surface, the uterus has a definite role in normal development. In response to the withdrawal of microvilli from the trophectoderm surface, the uterine cell surfaces also flatten and then develop some bulbous cytoplasmic projections for a limited period. With the loss of uterine sensitivity, these projections disappear and the epithelial cell surfaces acquire the microvilli once again. These changes are under hormonal control. Differences between the microvilli of implantation chambers and other parts of the uterus have been reported.[82] Freeze-fracture studies intended to determine the number and density of intramembrane particles have been reported. An increase in the density of intramembrane particles has been found to coincide with the time of implantation. This suggests an increased fluidity of the surface, which may facilitate relocation of epithelial receptors of various signals. After the positive role of the uterine epithelial cells is over, they are removed by a self-destructive process. The epithelial cells seem to be programed to die under a proper stimulus, even in the absence of a blastocyst.[86]

In certain animals such as the pig, development occurs entirely within the uterine lumen. The trophoblast does not invade the uterine epithelium and stroma. In ectopic sites, however, the pig trophoblast is capable of exhibiting invasive properties. In other species where implantation is deeper, the trophoblastic cells show remarkable invasive properties. They penetrate between the uterine epithelial cells across the basement membrane and subsequently deep into the stroma. The deep invasion is due to proteolytic activity as shown by the fact that protease inhibitors can prevent mouse implantation.[87] A noteworthy enzyme participating in this process is the plasminogen activator, an enzyme having the ability to convert the zymogen plasminogen to plasmin. Such an enzyme occurs in the trophoblast cells of mice and other animals. It has been suggested that the enzyme facilitates invasion of the uterine tissue by the trophoblast. Axelrod[88] has studied an implantation-defective mutant in mice. The homozygous recessive genotype fails to implant though the embryo proper is normal,

Table 3
LECTIN BINDING TO PREIMPLANTATION MOUSE BLASTOCYSTS

Lectin	Sugar specificity	Binding to blastocysts[a]	
		Day 5	Day 6
Dolichos biflorus agglutinin (DBA)	N-acetyl D-galactosamine	− (6)	− (6)
Soybean agglutinin (SBA)	N-acetyl α-D-galactosamine; D-galactose	− (6)	− (6)
Ulex europeus agglutinin (UEA)	α-L-fucose	+ (6)	+ (6)
Concanavalin A (Con A)	α-D-glucose; α-D-mannose	+ (6)	+ (6)
Wheat germ agglutinin (WGA)	N-acetyl-β-(1→4)-D-glucosamine; sialic acid	+ (6)	+ (6)
Ricinus communis agglutinin (RCA I)	β-D-galactose	+ (12)	−/+ (3/9)
Peanut agglutinin (PNA)	D-gal-β-(1→3)-acetyl galactosamine	− (12)	+ (12)

[a] Number of blastocysts in parentheses.

From Chavez, D. J. and Enders, A. C., *Dev. Biol.*, 87, 267, 1981. With permission.

albeit somewhat retarded. Experiments involving transplantation of such blastocysts into testis sacs of adult mice suggest that the homozygous mutant trophoblasts are deficient in invasive properties, and this is linked with low plasminogen.

The uterine epithelial cells presumably transmit some stimulus from the blastocyst to the deeper tissues where a number of changes take place. One of the changes thus induced is what is called the decidual reaction. Decidual cells are those that are shed at parturition. They arise from the uterine stroma fibroblasts. They have two or more nuclei that are polyploid, with more than 64 times the haploid DNA content. Their multinuclear condition is due to endoreduplication and not cell fusion. These cells are characterized by very high alkaline phosphatase content. They also contain glycogen and some fat. The decidual cells are likely to provide nutrition to the embryo. Another noteworthy change in the uterine stroma is an increased vascularization and enhanced permeability of the capillaries. This can be demonstrated by the injection of a dye of large molecular weight such as Pontamine sky blue or Giegy blue, intravenously. The dye appears clearly "extravasated" at implantation sites.

In most of the mammals studied so far, the positioning of the blastocyst with reference to the uterine anatomy is constant.[89] So also is the part of the blastocyst that makes the first contact with the uterine wall. A specific orientation of the blastocyst exists within the uterine horn. Studying implantation in the mouse, Smith[90] found that the uterine wall reacts to the presence of a blastocyst in contact. The abembryonic pole of the blastocyst first contacts the right or left wall at random. The formation of the implantation chambers then proceeds in a characteristic "mirror image" fashion on the two opposite sides.

The brief account of implantation given above is enough to indicate clearly that this is one of the most intricate developmental adaptations in which two distinct organisms show their respective changes which are perfectly timed and sequenced so as to constitute an harmonious single developmental process. It is also clear that these changes are, to a large extent, those associated with the cell surfaces. There is sufficient evidence to show that some system of cell recognition through surface alterations is at work. Thus the avenue for further investigations is now well defined.

V. CELL INTERACTIONS IN THE DEVELOPMENT OF TISSUES AND ORGANS

Cell differentiation is primarily the synthesis of products that characterize one cell as distinct from another. After the synthetic process has progressed considerably, gross morphological differences appear. For an embryologist, whether a cell has differentiated or not is evident from the presence or absence of some morphological differences often expressed as changed shape or altered behavior. The changes occurring in individual cells result in the alteration of the overall shape of a group of cells taking them into a higher level of tissue organization. Production of extracellular material and its organization into specific supramolecular arrangements is an additional aspect of this process. For conceptual simplicity, almost all developmental changes can be visualized as the programed synthesis of cytoplasmic and extracellular products and the consequent change of shape of the cells with their continuous interaction among themselves and with the external signals. Such changes are depicted in Figure 20, which represents two different generalized schemes of tissue and organ differentiation. There are many examples of development wherein the organ primordium (the first recognizable, histologically undifferentiated rudiment of an organ) is in the form of a sheet of cells. Regional differences in the multiplication of the component cells in an epithelial sheet generally result in one of the following possible transformations.

1.   If the cells have a supportive base, say in the form of accumulated mesenchymal cells

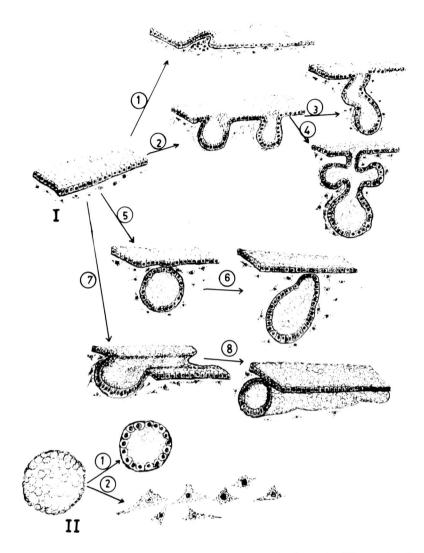

FIGURE 20. Diagram representing morphogenetic changes leading to the differentiation of tissues. I. The changes undergone by an epithelial sheet; (1) development into a structure resembling a feather/scale rudiment, (2 to 4) development into invaginated bending or branching tubes such as glands; (5 to 6) development of invaginated cells into a vesicle, completely separated from the superficial epithelial layer, and (7 to 8) depression over a long stretch of the epithelium forming a groove and eventually a tube. II. A mass of cells developing a cavity (1), or dispersing into discrete cells (2.)

and their extracellular products, a small raised region in the form of a miniature hillock will be formed. This is how the feather germs and scale rudiments develop. In animal embryos, there are characteristic spatial patterns in which these structures develop in the different anatomical regions.

2. If, on the other hand, the underlying connective tissue "gives way", the result is a localized depression that may subsequently expand in various ways. If it elongates, it takes the form of a tubular gland, the original connection with the surface representing the duct. Various patterns of coiling or branching of the tubular structure are additional features that may develop in the glands.

3. The localized depression may, alternatively, assume the form of a vesicle, which may eventually pinch off from the surface epithelium and thus be surrounded by the con-

Table 4
DIFFERENTIATION OF CHICK EMBRYONIC ECTODERMAL AND ENDODERMAL STRUCTURES IN RESPONSE TO THE INDUCTIVE INFLUENCE OF MESODERM

Interacting tissues	Typical result	Ref.
Dorsal dermis + epidermis of tarsometatarsal area	Feathers	91
Dorsal epidermis + dermis of tarsometatarsal region	Scales	91
Lung bud endoderm + stomach mesoderm	Gastric glands	92, 93
Lung bud endoderm + intestinal mesoderm	Villi	92, 93
Lung bud endoderm + bronchial mesoderm	Bronchial buds (branching)	92, 93
Lung bud endoderm + tracheal mesoderm	No branching	92, 93

nective tissue. Various regional patterns of growth in the vesicle can lead to diverse structures. The lens and the auditory vesicle of the vertebrates may be mentioned as typical examples. In case of the lens, a zone of cell proliferation adds new cells to one half of the vesicle where lens fibers differentiate. Complex deformations undergone by the auditory vesicle eventually result in the development of the membranous labyrinth.

4.   If the cells of the epithelial rudiment exist as an extensive plate and grow in a long narrow area, they can sink into the connective tissue, forming a long furrow. Eventually, when the edges of the furrow come closer and fuse, a tube is formed. This is how the central nervous system of the vertebrates develops. Regional differentiation along the neural tube gives rise to the brain with its subdivisions, and the spinal cord.

Morphogenetic organization can also start from a cellular mass. The development of a cavity within the mass leads to the formation of an epithelialized tissue, partly or wholly surrounding the cavity. In the development of the vertebrate somites, a cavity (myocoel) arises in the center of the cellular mass by a process akin to this. Cells may begin to disperse from a cellular mass. This is how the sclerotome of the somites behaves. The cells may move to a different location, aggregate around another structure, or penetrate it.

Most of the organs in the body consist of tissues derived from two of the three germ layers. Thus there are organs composed of ectoderm and mesoderm, or endoderm and mesoderm. Well defined anatomical structures composed of ectoderm or endoderm only do not exist. The vertebrate eye lens may be mentioned as a notable exception to this general rule. In most "endodermal" or "ectodermal" organs, however, there is always a varying proportion of connective tissue and vascular structures penetrating them. During development, there is a continuous interaction between the two components of the organ rudiments as shown by experimental embryologists. An essential aspect of the interaction is that the developmental program of one component is influenced and defined by the other. A variety of ectodermal structures, which are characteristic of particular body regions, are determined through an interaction with the underlying mesoderm. Similarly, endodermal differentiation into a variety of structures is determined by the mesoderm interacting with it (Table 4). A striking feature that emerges from these studies is that the mesodermal component is essential. Besides, in many cases, the ectodermal or endodermal differentiation is appropriate to the mesodermal component. In spite of the large number of instances of such epithelio-mesenchymal interactions, no molecular or cellular mechanism has so far been elucidated in any case. The interactions are of a very complex nature.

A. Neurulation

The development of the central nervous system in the vertebrates is a complex morphogenetic process. It is known that the pattern in which the different homologous brain divisions

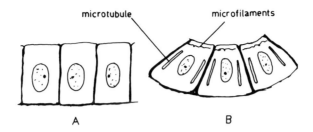

microtubule   microfilaments

A   B

FIGURE 21.   Transformation of columnar epithelial cells into wedge-
shaped cells in the development of a tubular structure from an epithelial
sheet. Note the location of microfilaments and microtubules. Contraction
of microfilaments is suggested as a mechanism bringing about the "purse
string" effect.

are formed is remarkably constant in all the vertebrates. During the early development of
the brain and spinal cord, one may therefore assume that similar mechanisms operate.
Obviously, these mechanisms must have been conserved during evolution. A detailed study
of the process, especially with reference to its molecular and cytological basis, would be
highly rewarding. The histological changes that occur during neurulation are fairly well
studied in all the vertebrates. A thickening of the dorsal ectoderm is the earliest visible
change. The cells overlying the notochord become columnar. Localized cell proliferation is
one of the causes of thickening. However, to assume and maintain the columnar shape, the
cytoskeletal organization and the surface characteristics of the cells, especially where they
are adhering to each other, must be highly specialized. The presence of tight junctions
characterizes many epithelia. It is generally held that the tight junctions are responsible for
maintaining the epithelial polarity. During neurulation in *Rana pipiens* embryos, specific
rearrangement in the organization of tight junctions occurs,[94] which may well be a general
feature of all vertebrates. There is a disassembly of these membrane specializations as
neurulation progresses. A redistribution of intramembrane particles between the apical and
lateral surfaces of the epithelial cells has been found to accompany the disassembly of the
tight junctions.[94] These changes are probably related to the alterations in membrane polarity,
which occur as neurulation progresses. A striking feature of the neural plate is the presence
of the slightly raised margins on either side. These are called neural folds. From their
subsequent development into the neural crest, it is clear that they must be quite different
from the neuroepithelium that eventually constitutes the neural tube.

As the neural plate transforms into a neural groove and eventually into a tube, a complex
set of morphogenetic changes occurs. For a long time it has been recognized that an important
aspect of the mechanism of neurulation is a characteristic and highly organized change in
the shape of the component cells, as first suggested by Gläser.[95] A definite correlation
between the organization of the cytoskeletal elements and the changed cell shape was
suggested by Burnside[96] and Schroeder.[97] A contraction of the subcortical microfilaments,
which are distributed horizontally at the cell apex (i.e., parallel to the luminal surface),
would result in its narrowing in the manner of a purse string. Elongation of the cytoplasmic
microtubules would complement this action (Figure 21). Recent studies have made use of
scanning electron microscopy of fractured neuroepithelia to reveal the change in the shape
of cells. These studies, supplemented by transmission electron microscopy, have helped
correlate the changes of cell shape with the organization of the cytoskeletal elements.[98]
Disruption of microfilaments by cytochalasin B or of microtubules by colchicine results in
the development of highly abnormal neural tubes in amphibian and avian embryos. Generally
the neural plates of such embryos fail to close into tubes. Considerable work has been done
on the teratogenic action of certain chemicals which in general interfere with the oxidation-

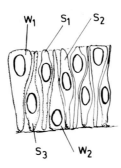

FIGURE 22.   Diagram to show the different shapes of cells in the bending neural plate. W_1 and W_2 are different wedge shapes and S_1, S_2, and S_3 are different spindle shapes. See the text.

reduction of sulfhydryl (−SH) groups.[99-101] The actions of these chemicals on the developing embryos are different, but the syndromes are similar and reminiscent of those caused by agents that interfere with the polymerization of the cytoskeletal structures. On the basis of these observations, it is difficult to infer any molecular changes leading to alteration in the shape of cells during normal development. Such studies can, however, lead to the identification of a common lesion caused by the different chemicals. It is also clear that the mechanism of neurulation consists of coordinated changes both at the cell surface and the cytoskeleton.

A simple pattern of the change of cell shapes in the transformation of a plate into a tube can be suggested. If the cells change their cuboidal or columnar shape and assume a wedge shape while continuing to retain their adhesion to the neighboring cells, a tubular structure will be formed. Attempts have been made to analyze the shape changes mathematically on the basis of this simple assumption.[102,103] However, not all neuroepithelial cells assume the wedge shape uniformly. The pattern in which they change their shape varies across the neural plate. Bending of the neural plate takes places in three distinct locations: a median-ventral (supranotochordal) and two dorso-lateral. It is in these regions that significant changes take place in the shape of cells. Here many cells become wedge shaped, and this seems to be correlated with the bending of the neural plate.

Schoenwolf and Franks[98] have analyzed statistically the data on cell shape changes in the neural tube and found that a large number of cells are wedge shaped in these regions. In the intermediate regions, the wedge-shaped cells are less numerous and the differences are found to be statistically significant. In the neural tube, not all cells are wedge shaped. There are two broad classes of shape exhibited by these cells: wedge and spindle. The former may be oriented with the broad end toward the apex or the basement membrane.[98] Evidently, these two subtypes are formed by opposite mechanisms as suggested by the fact that in the first case it is the apical (luminal) surface that has to expand and, in the other, it is the surface apposed to the basement membrane that is extended. The early neural tube is a pseudostratified epithelium with cells of these shapes fitting with each other so as to assume a fairly compact arrangement (Figure 22). A definite program of change in the shape of cells presumably exists in the neurulation process. Attempts at a mathematical analysis of such a program will have to take into account not only the two shapes of cells but even small variations around the idealized spindle and wedge shapes. As shown in Figure 22, there could be more than one spindle shape. It is easy to show that the possible number of spindle shapes is determined by the size of the nucleus, its deformability, and the thickness of the pseudostratified epithelium.

It will be instructive to pose the question as to whether the specific pattern of shapes is the effect rather than the cause of neurulation. In case it is the effect, one needs to look for the cause, which must precede in time. It is conceivable that extrinsic forces can cause deformations in the neuroepithelial cells. The suggested extrinsic causes include medial migration of the nonneural epithelium and influences of the notochord and extracellular matrix. Neurulation occurs in the prechordal region also. It seems therefore that the notochord does not initiate the changes in cell shape. Schoenwolf and Franks[98] suggest that mitotic activity in the neural plate may also contribute to the process of cell deformations and the consequent folding. It is known that intermitotic displacement of nuclei occurs in the neuroepithelial cells,[104] and the position of the nucleus in a cell could determine its shape. Thus a definite spatial plan of cell divisions could contribute to the process of neurulation.

Mutual adhesiveness of cells has been considered to be another component of the neurulation mechanism. *A priori*, it may be assumed that a change from cuboidal to columnar shape can be brought about by an increase in the mutual adhesiveness of the cells. Brown et al.[105] considered that this could be the mechanism by which the neuroepithelial cells become columnar. The idea finds the support of recent workers also. Childress,[106] discussing the geometrical aspects of changing cell shape, has emphasized the role of increased cell adhesiveness during neurulation. Direct estimations of adhesiveness have, however, shown that cells become less adhesive in the neuroectoderm compared with those in the nonneural ectoderm.[107] Though an increase in the adhesiveness of neural cells was found during development, the difference still remained, as the nonneural ectodermal cells also acquired increased adhesiveness during the same time. On balance, it may be said that some of the work on this problem deserves repetition and reconfirmation using improved methods. In particular, since considerable progress has been made on isolation and characterization of cell adhesion molecules (Volume I, Chapter 2), it would be worth directing renewed attention to the question of changing adhesiveness of the neuroepithelial cells during development.

B. Coiling and Branching of Tubular Structures

The development of a tubular structure can start from an epithelial primordium or from a mass of cells. In the latter case, a cavity develops, giving the surrounding cells an epithelial character. This is how the kidney tubules are formed. The reverse of this process, i.e., loss of epithelial character and assumption of mesenchymal features, is also known to occur as in the formation of the neural crest (Chapter 3). Once a tube is formed it elongates, dilates locally, or its walls thicken, thus diversifying it morphologically. The basis of these alterations probably consists of such cellular activities as change of shape, mutual adhesiveness, and interaction with the cellular and noncellular milieu. The amazing variety of tubular structures developed in embryos, however, suggests that these cellular activities are modified variously so as to control the development of distinct types of tubes. If a thin tubular structure grows by the multiplication of cells that retain their cuboidal shape, the result will be a continued lengthening of the tube, and if it is subjected to mechanical or other constraints by the surrounding material, bending or coiling will occur. Many tubular structures exhibit this pattern of morphogenesis. The bending and coiling is highly patterned. The characteristic convolutions and loops of vertebrate kidney tubules are developed into a definite pattern with reference to each other and to the vascular supply.

The control mechanisms of these developmental changes are obscure. Earthworms have a complicated osmoregulatory and excretory system which consists of highly coiled tubules known as nephridia. There are some species in which the nephridia are paired in each segment; there are others in which a large number of nephridia occur in each segment. Further, their pattern of coiling is characteristic.[108] How the developmental mechanism is controlled in the formation of these tubular coiled structures is not known. An additional point for further investigation is how a group of cells, which form a single nephridium in some species, can split and produce a clone of 100 or more such structures in others. The amazing variety of tubular structures developed in animal embryos provides ample experi-

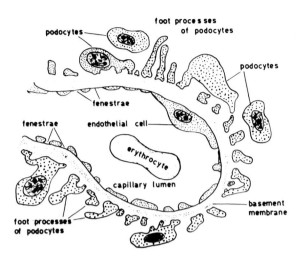

FIGURE 23.   The location of podocytes in the kidney glomerulus. Note
the characteristic shapes of the podocytes and their foot processes. The
capillary endothelial cells are fenestrated. See the text.

mental material to investigate the cellular and molecular basis of growth, coiling, and other
aspects of differentiation.

Tubular structures contain different types of cells in their epithelia. Besides the cuboidal
and columnar shapes that occur most frequently, a curious cell type must be mentioned.
The epithelial cells of the kidney glomerulus are remarkable for their unusual shape. These
cells, known as podocytes, are in contact with the basement membrane only at the ends of
their foot-like extensions, which leave considerable free area for direct filtration of the
primary filtrate (Figure 23). Maintenance of their shape is vital to their function. A number
of kidney disorders are correlated with abnormal podocytes. What maintains their shape is
not fully understood. Their surfaces are highly negatively charged, owing to the presence
of a surface coat called epithelial polyanion, which is rich in sialic acid. Several podocyte
surface antigens have been identified.[109] One of them, the major sialoprotein component, is
called podocalyxin.[110] The surface components are presumably important in maintaining the
cell shape as indicated by a correlation between certain glomerular diseases and the absence
of stainable surface material.[111] Infusion of polycationic compounds causes disorganization
of podocyte morphology.[112] All these facts suggest a definite role for the surface sialoproteins
in the maintenance of podocyte morphology and function.

Several of the embryonic tubular structures develop as highly branched epithelia. The
epithelial component in these structures may be endodermal (e.g., the lung) or ectodermal
(e.g., the mammary gland and salivary gland). During their differentiation, a repeated
branching occurs so as to give rise to a system of secretory ends connected by ducts.
Proliferation of cells and changes in their shape seem to be the most obvious mechanisms
of tubular morphogenesis.[113] Goldin[114] has reviewed the literature on this subject and brought
forward the following sequence of events in the development of branched tubular organs.

1.   The basement membrane of the epithelium is degraded at the point where growth of
the epithelium is programed to occur.
2.   The above event facilitates establishment of a close (or direct) contact between the
epithelium and mesenchyme. Stimulation for localized cell proliferation at this point
passes from the mesenchyme to the epithelium.
3.   Contraction of microfilament bundles facilitates bulging of the part where proliferation
occurs.

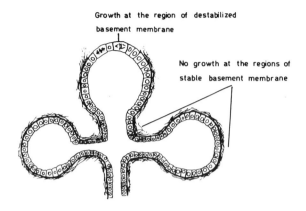

Growth at the region of destabilized basement membrane

No growth at the regions of stable basement membrane

FIGURE 24.   Changes in the extracellular matrix during branching morphogenesis.

4.   Accumulation of collagen fibrils and formation of a definite basement membrane occur at the regions where the tube is not growing.

There seems to be little doubt that a fibrillar collagen network and a stable basement membrane are characteristic of the regions of epithelia where morphogenetic activity is absent. In contrast, regions of growth are characterized by the absence of fibrillar collagen and a firm basement membrane (Figure 24). Several studies have indicated that the glycosaminoglycan component of the basement membrane has a definite role in the branching morphogenesis of several types of mouse epithelia.[115-117] There is a rapid turnover of hyaluronate and chondroitin sulfate at the active sites of branching, while these substances accumulate at the quiescent clefts between the branches. When these substances are removed enzymically from the epithelium and then cultured in combination with mesenchyme, branching is inhibited. Untreated epithelium combined with mesenchyme continues to exhibit branching in culture. If the enzyme-treated epithelium is preincubated for 2 hr to restore the basement membrane and then combined with mesenchyme, branching continues. Interfering with the biosynthesis of xylose-linked classes of glycosaminoglycans also leads to inhibition of branching of mouse salivary gland in vitro. Thompson and Spooner[118] used β-D-xyloside as a tool to inhibit the normal glycosylation process, which consists of xylosylation of certain serine residues of the core protein followed by the addition of two galactose residues. Further addition of monosaccharides then proceeds. When β-D-xyloside is present, it competes with the serine-bound xyloside and leads to the synthesis of free glycosaminoglycans. Treatment of the salivary glands in organ culture at different time intervals (Table 5) showed that a continued interference with the normal synthesis of glycosaminoglycans results in the inhibition of branching.

It is interesting to note that the effect is reversible as shown by normal morphogenesis in cultures exposed to the inhibitor only for the initial 24 hr period. The turnover of the basement membrane requires the synthesis of Type IV collagen also. Inhibition of the synthesis of this component of the basement membrane by treatment with *cis*-hydroxyproline, a proline analogue, prevents the growth and branching of rat mammary gland in vitro as well as in vivo.[119] Spooner and Faubian,[120] using L-azetidine-2-carboxylic acid (another proline analogue), have demonstrated inhibition of branching in lung and salivary gland tissue. A general conclusion that emerges from the above observations is that a controlled regulation of the turnover of basement membrane provides a flexible basement membrane in the regions where branching occurs.[121]

Table 5
BRANCHING MORPHOGENESIS IN MOUSE
SALIVARY GLANDS FOLLOWING
CULTURE IN β-D-XYLOSIDE

Culture Conditions[a]			
0—24 hr	24—48 hr	48—72 hr	Rate of branching[b]
Nil	Nil	Nil	4.33 ± 0.28 (35)
0.5 m*M*	0.5 m*M*	0.5 m*M*	0.88 ± 0.09 (14)
1.0 m*M*	1.0 m*M*	1.0 m*M*	0.73 ± 0.08 (16)
0.5 m*M*	Nil	Nil	5.13 ± 0.49 (28)
1.0 m*M*	Nil	Nil	4.32 ± 0.61 (17)

[a]   Concentration of β-D-Xyloside in the medium during the 72 hr
       culture is shown.
[b]   The rate of branching is expressed as mean number of lobes at
       72 hr per number of lobes at 24 hr ± SEM (N).

From Thompson, H. A. and Spooner, B. S., *Dev. Biol.*, 89, 417,
1982. With permission.

C. Segmental Organization of the Body

A very broad generalization in the field of comparative anatomy is that in many animal
groups the body plan is organized on a pattern of repeating units from the anterior to the
posterior end. The repeated units are known as segments or metameres. Whether the metameric pattern arose only once or several times during evolution has been debated. All
chordates are segmented. Among the invertebrates, Annelida and Arthropoda are typically
segmented. There are other minor groups which have a segmented body. Among the Mol-
lusca, an almost extinct group of Monoplacophora, now represented by the genus *Neopilina*,
possess a metameric body organization. This probably links some early mollusks with the
two other major invertebrate phyla that exhibit segmentation. It will not be profitable for
us to dwell upon the question of evolutionary continuity of the segmental pattern. However,
it is pertinent to emphasize that segmentation is a well-established developmental process.
It seems reasonable to assume that the basic mechanisms of mesodermal segmentation during
the development of chordates have been conserved during evolution. Another interesting
feature of the metameric pattern is its regional differentiation along the body axis. In all
probability, the regional differences arise due to minor alterations of the basic mechanism.
Elucidating the basic mechanism of segmentation and its alteration leading to the differentiation along the body axis therefore constitutes a fundamental problem of developmental
biology.

For a given chordate species, the number of segments is remarkably constant. Generally
there is a craniocaudal sequence in which the somites are differentiated so that any somite
is ahead in the developmental schedule compared with the one behind it.

Attempts have been made to elucidate somitogenesis by following different avenues of
approach, which may be considered under three different categories. First, descriptive investigations to reveal the sequence of changes taking place in the formation of somites have
been pursued using different methods. Information obtained from the classical histological
studies has been supplemented by transmission and scanning electron microscopy. In particular, the latter method has yielded a wealth of information on the shape changes undergone
by the cells. The descriptive approach has been extended to comparing the adhesive and
other properties of the cells before and after segmentation. Since somite formation proceeds
cranio-caudally, tissues passing through different phases of the process are readily available

FIGURE 25. Scanning electron micrograph of the dorsal view of a longitudinally fractured *Xenopus* embryo illustrating the changing orientation of myogenic cells in the differentiating somites. Two segmented somites and the posterior unsegmented mesoderm are shown. (From Youn, B. W. and Malacinski, G. M., Somitogenesis in the amphibian *Xenopus laevis*. Scanning electron microscopic analysis of intra-somitic cellular arrangements during somite rotation, *J. Embryol. Exp. Morphol.*, 64, 23, 1981. With permission of the publisher, the Company of Biologists, Cambridge University Press.)

from the same embryo. Extracellular material associated with the somitogenic mesoderm has also been studied. The second approach is the use of classical methods of experimental embryology. These include disturbing the relation between the paraxial somitogenic mesoderm and the neighboring structures such as the neural tube, notochord, and endoderm[122,123] and other operative procedures on the early stages of embryonic development.[124-126] Besides, chemical methods of interfering with the synthesis of various substances during the process have been used. Any abnormality following such treatments is correlated with a role for the substances that are affected. Finally, several workers[127,128] have approached the problem at the theoretical level, with reference to the temporal and spatial aspects of somite development.

Somites are developed from the vertebral plate (paraxial mesoderm), which is initially unsegmented. A block of cells is separated by the formation of visible boundaries and begins to assume a rosette-like arrangement. In *Ambystoma* embryos, Youn and Malacinski[129] have shown that the vertebral plate of the mesoderm is first organized as a continuous band of cells without showing any visible boundaries of somites. It consists of vertically elongated cells in two tiers. Gradually the boundaries of somites become visible (Figure 25). A series of shape changes ensues in the group of cells segregated as a somite. The cells at the boundaries assume a somewhat triangular shape and thereby accentuate the rosette arrangement.

This pattern of segmentation has been modified in the anurans, particularly in *Xenopus*. Hamilton[130] described a striking feature of cell movement and orientation during the formation of somites in *Xenopus laevis* embryos. It may be noted that in this species, overt segmentation occurs only in the myotome; the dermatome remains unsegmented. After the segment has become defined, the cells destined to form the myotome rotate so that their medial ends point anteriorly. Somitogenesis has been studied in the embryos of higher vertebrates also. In the chick embryo, the vertebral plate is first recognizable as a thick band of mesodermal cells extending longitudinally beside the neural tube and notochord. Initially it consists of loosely arranged columnar cells. Later, the cells assume a more compact organization. Groups of loosely packed cells arranged as an epithelium first segregate as a unit, constituting a somite. The cells are then arranged as an epithelium around a small space, the myocoel, where they are joined by tight junctions. In the myocoel are found some cells of an irregular shape. At the opposite end, i.e., the basal, the cells of the rosette have a basement membrane. Nearer the basal end, the cells are attached to one another by desmosomes. Some of the cells in the rosette are wedge shaped, whereas the others are spindle shaped.[131] Subsequent

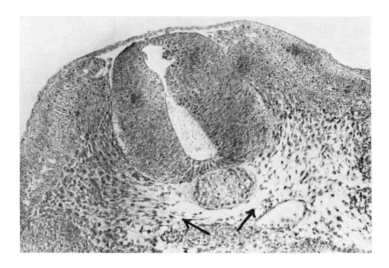

FIGURE 26.   Cross section of a chick embryo (about 80 hr) showing the cells of
the sclerotome (arrows). See the text.

differentiation of the somite progresses when the dermatome, myotome, and sclerotome
arise from the rosette of cells. The dorsolateral cells of the rosette remain as a columnar
epithelium with each cell in it spanning the thickness. These cells constitute the dermatome.
The myotome, constituted by the dorso-medial border of the rosette, gradually extends to
reach the lateral edge of the dermatome, along the inner margin. The myocoel is now almost
obliterated and the composite plate formed is often called the dermomyotome. The ventro-
lateral cells of the rosette, constituting the sclerotome, begin to disperse by assuming irregular
shapes (Figure 26).

When we turn to considering the role of the cell surface in somitogenesis, it is found that
surprisingly little is known. Most of the experimental work has been addressed to the question
of determination of the somite number, the prospective significance of the vertebral plate,
the possible "inducing" influence of other neighboring structures, the "regulative" powers
of the vertebral plate, and other related problems. These efforts have presented the problem
in a clearer perspective to make a more direct experimental approach to elucidate the role
of the cell surface.

Considerable differentiation at the molecular level presumably occurs even before a group
of cells in the undifferentiated vertebral plate assumes the segmental boundaries. Some
particular aspect of this early differentiation is heat shock sensitive.[132] The earlier and later
phases are not sensitive. This is concluded from the observation that in *Xenopus* embryos
exposed to a heat shock of short duration (up to 15 min) some somites (4 to 6) continue to
form normally, presumably from the tissue which has become refractory to the treatment
after having passed through the sensitive phase. A few (2 to 6) abnormal somites are then
formed, presumably from the tissue which, at the time of administering the heat shock, was
in the sensitive phase. Subsequently, normal somites begin to form from the tissue that had
not yet passed through the sensitive phase of differentiation. Ballairs et al.[107] correlated
elongation and an increase in the mutual adhesiveness of cells during the formation of
segments. They have, using the couette viscometer technique, shown that the collision
efficiency of the cells from the segmented mesoderm is about twice that of the cells from
the unsegmented vertebral plate. It has been suggested that the elongation of the cells and
a localized increase in their mutual adhesiveness at the apical ends are the principal causes
of rosette formation in chick somitogenesis. This conclusion is supported by the observation
that trypsinized rosette cells become rounded and have no organized microfilament bundles,
unlike the ones in the elongated form.

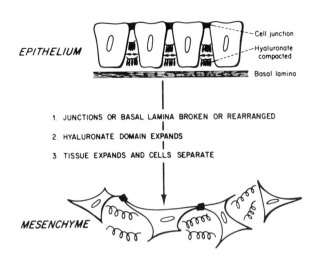

FIGURE 27.   A model for expansion of intercellular spaces. Prior to expansion, hyaluronate is secreted between epithelial cells and held in a compact configuration due to confinement by cellular junctions or the basal lamina. On breakage or rearrangement of these restraints, the swelling pressure exerted by the hyaluronate meshwork (represented by the tight coiled springs) causes expansion of the intercellular spaces and separation of the cells. (From Toole, B. P., *Cell Biology of Extracellular Matrix,* Hay, E. D., Ed., Plenum Press, New York, 1981, chap. 9. With permission.)

Evidence for a definite role for increasing cell adhesiveness in reorganizing the cells during somite formation has been provided by the recent work of Cheney and Lash.[133] It has been suggested that the cells in the rosette are under tension like the fibroblasts, which have stretched out in vitro. Fine collagen fibers extending between the rosette cells and the neighboring structures (e.g., ectoderm, endoderm, aorta, and lateral plate), have been observed in scanning electron micrographs.[131] A positive role for the collagen fibers has been demonstrated.[125] Administration of α-α-dipyridyl, which is an iron chelator and affects the secretion of collagen, exerts definite effects on chick somitogenesis. The chemical prevents the segmentation of the vertebral plate, and the susceptible region moves posteriad as development proceeds. Fibronectin also has some role in the process of compacting the somitogenic cells.[134]

The migration of sclerotomal cells from the rosette shaped rudiment must be a process aided by extracellular matrix components. Considerable evidence has been accumulated indicating that it is so. Kvist and Finnegan[135,136] demonstrated that the migrating cells are surrounded by hyaluronate. In contrast, around the notochord where cartilage forms, it is chondroitin sulfate and not hyaluronate that is the abundant matrix material. The mesenchymal cells migrate through a hyaluronate-rich matrix. The leading edge of the migrating cell, in particular, is surrounded by the proteoglycan granules and filaments as revealed by ruthenium red staining.[137] It is possible that the migrating mesenchymal cells have surface receptors for binding and migrating through the highly hydrated hyaluronate-rich matrix (for a general discussion on the matrix-cell surface interactions in different morphogenetic processes, see the reviews by Toole[121] and Hay.[138] Embryos treated with hyaluronidase show the failure of sclerotomal cells to separate and disperse.[139] This indicates clearly that the hydrated space offered by the hyaluronate-rich matrix facilitates the migration of cells.

It may be noted that the sclerotome of the somite is present as an epithelial layer in the rosette-shaped somite. The transformation of epithelial cells into free mesenchymal cells necessarily involves some important changes at the cell surface. Toole[121] has outlined some of the essential changes in an epithelium that should take place leading to such a transformation (Figure 27). Hyaluronate held compactly between epithelial cells may expand when its hydration is facilitated by a rearrangement of junctional complexes and changes in the

Table 6
SEGMENTATION OF THE INSECT BODY

Segment	Appendages
Head	
1. Preantennal	Absent
2. Antennal	Antennae
3. Premandibular (intercalary)	Embryonic
4. Mandibular	Mandibles
5. Maxillary	Maxillae
6. Labial	Labium
Thorax	
1.Prothorax	1st pair of walking legs
2. Mesothorax	2nd pair of walking legs, 1st pair of wings
3. Metathorax	3rd pair of walking legs, 2nd pair of wings
Abdomen	
Abdominal segments I—X	Embryonic

Note: Insects belonging to different orders and families show considerable variation in the adult structure. The appendages are also variously modified. For a detailed discussion on the head segmentation, see Du Porte,[142] Anderson,[143] and Jura.[144]

composition of the basement membrane. In vitro studies[140] have shown that the environmental conditions can have a profound effect on the epithelial phenotype. Epithelial cells suspended in three-dimensional gels of native collagen elongate, detach from the explant, develop pseudopodia, and migrate as individual cells. One wonders if the collagen between the notochord and the somite has any role in the dispersal of the sclerotomal cells. It is quite possible that there are additional factors working in unison to bring about the highly patterned migration of the sclerotomal cells. The information available so far is perhaps incomplete. Nonetheless, it clearly indicates the direction of further approach for a fuller elucidation of the process.

Some recent research in another area of biology, viz., genetics, is likely to take us far ahead in understanding the segmental organization of animal bodies. Insect segmentation has been studied by comparative anatomists and embryologists for a long time. The segmental organization of the insect head and the question of homologous structures in other arthropods have been the subject of lively discussion in the past. The segmental body plan of an insect such as the fruit fly, *Drosophila melanogaster,* is depicted in Table 6. From the table, it is clear that some anterior body segments are characterized by the presence of different types of appendages.

Studies on the genetics of *D. melanogaster* have now opened a new avenue of approach to the problem of segmentation and, in particular, to the regional differentiation of the somites. Mutations affecting the differentiation of segmental structures were discovered about 70 years ago.[141] Recent work has revealed the existence of at least two gene complexes located on the right arm of the third chromosome, which affect early development of the segments. One of them, the bithorax complex, is concerned with the differentiation of the posterior segments. The other, the antennapedia complex, controls the anterior region. A normal fruit fly, like all other diptera, has a pair of wings on the mesothorax and a pair of halteres or balancers on the metathorax. There are mutations in the bithorax complex that affect the development of the third thoracic segment leading to the formation of structures

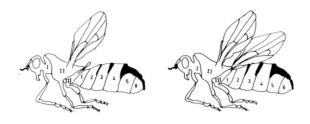

FIGURE 28.   The bithorax complex of *Drosophila melanogaster*. In the normal fruit fly (left) the middle thoracic segment bears a pair of wings and the third thoracic segment has a pair of halteres which are homologous to wings. In a phenotype which combines three mutations of the bithorax complex (right), two pairs of wings are developed. I to III, thoracic segments; 1 to 6, abdominal segments.

more akin to the wings. In fact, breeding the flies so as to combine three mutations into a single genotype produces individuals that have typical wing-like structures developed on the third thoracic segment (Figure 28). The genetic lesion for which the other complex was named "antennapedia" causes a transformation of the antenna to mesothoracic leg.[145] There is evidence to indicate that the two complexes interact.[146] Elegant breeding experiments aided by the recombinant DNA technology have succeeded in isolating the gene complexes. There are some surprising features in the genes of the bithorax complexes. For example, there are unusually long noncoding sequences in the DNA. There are also "genes" of which no translated products exist.

No doubt, the genetic basis of segmental differentiation must be a very complex process. Much of the genetic and embryological work[147] has resulted in revealing a definite relationship between the most fundamental mechanism (i.e., genes) and the developed anatomical parts. Recent experiments[145] have also helped to define the times and places of action of these genes during early development. No information is yet available on the causal relation between the genes and the structures controlled by them. It is, however, gratifying that considerable information on the genes is already available. Further investigations will undoubtedly give us a deeper insight into the mechanism of segmentation.

VI. CONCLUDING REMARKS

In an attempt to elucidate the cellular basis of embryonic development, we have selected a few striking examples that clearly indicate that some important components of the mechanism have been identified. A concept can be developed on the assumption that the cell surface has a central role to play in it. External signals (diffusible "inducing" molecules and hormones, and the extracellular matrix components) stimulate the cells in various ways. Neighboring cells also may communicate through junctional complexes. The synthetic activities of the cells are thus modified. These activities include not only the elaboration of intracellular substances but also the secretion of the matrix. Changes within the cytoplasm, especially those involving polymerization/depolymerization of actin and tubulin, are expressed as changes in cell shapes. Other overt features of differentiation will then follow. Thus a molecular interpretation of differentiation at the gross level is now feasible, at least in a broad outline.

REFERENCES

1. **Reverberi, G.,** Dentalium, in *Experimental Embryology of Marine and Freshwater Invertebrates,* Reverberi, G., Ed., North-Holland, Amsterdam, 1971, chap. 9.
2. **Ijiri, K. I.,** Existence of ultraviolet-labile germ cell determinant in unfertilized eggs of *Xenopus laevis* and its sensitivity, *Dev. Biol.,* 55, 206, 1977.
3. **Dictus, W. J. A., Van Zoelen, E. J. J., Tetteroo, R. A. T., Tertoolen, L. G. J., De Laat, S. W., and Bluemink, J. G.,** Lateral mobility of plasma membrane lipids in *Xenopus* eggs: regional differences related to animal/vegetal polarity become extreme upon fertilization, *Dev. Biol.,* 101, 201, 1984.
4. **Eddy, E. M.,** Germ plasm and differentiation of the germ cell line, *Int. Rev. Cytol.,* 43, 229, 1975.
5. **Brachet, J.,** An old enigma: the gray crescent of amphibian eggs, *Curr. Top. Dev. Biol.,* 11, 133, 1977.
6. **Ancel, P. and Vintemberger, P.,** Récherches sur le déterminisme de la symmétrie bilatérale dans l'oeuf de amphibiens, *Biol. Bull. Fr. Belg.,* 31, 1, 1948.
7. **Pasteels, J.,** The morphogenetic role of the cortex of the amphibian egg, *Adv. Morphogenesis,* 3, 83, 1964.
8. **Nieuwkoop, P. D.,** The "organization centre" of the amphibian embryo: its origin, spatial organization and morphogenetic action, *Adv. Morphogenesis,* 10, 1, 1973.
9. **Gerhart, J., Ubbels, G., Black, S., Hara, K., and Kirschner, M.,** A reinvestigation of the role of the grey crescent in axis formation in *Xenopus laevis, Nature (London),* 292, 511, 1981.
10. **Ubbels, G. A., Hara, K., Koster, C. H., and Kirschner, M. W.,** Evidence for a functional role of the cytoskeleton in determination of the dorsoventral axis in *Xenopus laevis* eggs, *J. Embryol. Exp. Morphol.,* 77, 15, 1983.
11. **Ubbels, G. A. and Brom, T. G.,** Cytoskeleton and gravity at work in the establishment of dorsoventral polarity in the egg of *Xenopus laevis, Adv. Space Res.,* 4 (12), 9, 1984.
12. **Kochav, S. and Eyal-Giladi, H.,** Bilateral symmetry in chick embryo, determination by gravity, *Science,* 171, 1027, 1971.
13. **Waddington, C. H.,** Experiments on the development of chick and duck embryos cultivated *in vitro, Philos. Trans. R. Soc. London,* B211, 179, 1932.
14. **Eyal-Giladi, H. and Fabian, B. C.,** Axis determination in uterine blastodiscs under changing spatial positions during the sensitive period for polarity, *Dev. Biol.,* 77, 228, 1980.
15. **Mitrani, E., Shimoni, Y., and Eyal-Giladi, H.,** Nature of the hypoblastic influence on chick embryo epiblast, *J. Embryol. Exp. Morphol.,* 75, 21, 1983.
16. **Stern, C. D.,** A simple model for early morphogenesis, *J. Theoret. Biol.,* 107, 229, 1984.
17. **Ubbels, G. A., Brom, T. G., Willemsen, H. P., and Van Nunen, J. J. H.,** The role of gravity in the establishment of the dorso-ventral axis in the developing amphibian embryo, in *Space Biology with Emphasis on Cell and Developmental Biology,* London, L. and Melital, O., Eds., ESA Scientific and Technical Publications, Noordwijk, 1983, 77.
18. **Brom, T. G., Ubbels, G. A., and Willemsen, H. P.,** An automatic device for amphibian egg fertilization in space: technical aspects and biological requirements, *Proc. 2nd Eur. Symp. Life Sci. Res. Space,* ESA Scientific and Technical Publications, Noordwijk, 1984, 13.
19. **Reverberi, G.,** Ascidians, in *Experimental Embryology of Marine and Freshwater Invertebrates,* Reverberi, G., Ed., North-Holland, Amsterdam, 1971, chap. 13.
20. **Speksnijder, J. E., Mulder, M. M., Dohmen, M. R., Hage, W. J., and Bluemink, J. G.,** Animal vegetal polarity in the plasma membrane of a molluscan egg: a quantitative freeze-fracture study, *Dev. Biol.,* 108, 38, 1985.
21. **Ducibella, T.,** Surface changes of the developing trophoblast cell, in *Development in Mammals,* Vol. 1, Johnson, M. H., Ed., North-Holland, Amsterdam, 1977, 5.
22. **Reeve, W. J. D. and Ziomek, C. A.,** Distribution of microvilli on dissociated blastomeres from mouse embryos: evidence for surface polarization at compaction, *J. Embryol. Exp. Morphol.,* 62, 339, 1981.
23. **Reeve, W. J. D.,** Cytoplasmic polarity develops at compaction in rat and mouse embryos, *J. Embryol. Exp. Morphol.,* 62, 351, 1981.
24. **Bilozur, M. and Powers, R. D.,** Two sites for calcium action in compaction of the mouse embryo, *Exp. Cell Res.,* 142, 39, 1982.
25. **Ducibella, T.,** Divalent antibodies to mouse embryonal carcinoma cells inhibit compaction in the mouse embryo, *Dev. Biol.,* 79, 356, 1980.
26. **Handyside, A. H.,** Distribution of antibody- and lectin-binding sites on dissociated blastomeres from mouse morulae: evidence for polarization at compaction, *J. Embryol. Exp. Morphol.,* 60, 99, 1980.
27. **Hyafil, F., Morello, D., Babinet, C., and Jacob, F.,** A cell surface glycoprotein involved in the compaction of embryonal carcinoma cells and cleavage stage embryos, *Cell,* 21, 927, 1980.
28. **Vestweber, D. and Kemler, R.,** Rabbit antiserum against a purified surface glycoprotein decompacts mouse pre-implantation embryos and reacts with specific adult tissues, *Exp. Cell Res.,* 152, 169, 1984.

29. **Shirayoshi, Y., Okada, T. S., and Takeichi, M.,** The calcium-dependent cell-cell adhesion system regulates inner cell mass formation and cell surface polarization in early mouse development, *Cell,* 35, 631, 1983.
30. **Bird, J. M. and Kimber, S. J.,** Oligosaccharides containing fucose linked α(1-3) and α(1-4) to *N*-acetylglucosamine cause decompaction of mouse morulae, *Dev. Biol.,* 104, 449, 1984.
31. **Kimber, S. J., Surani, M. A. H., and Barton, S. C.,** Interactions of blastomeres suggest changes in cell surface adhesiveness during the formation of inner cell mass and trophectoderm in preimplantation mouse embryo, *J. Embryol. Exp. Morphol.,* 70, 133, 1982.
32. **Surani, M. A. H. and Handyside, A. H.,** Reassortment of cells according to position in mouse morulae, *J. Exp. Zool.,* 225, 505, 1983.
33. **Pedersen, R. A. and Spindle, A. I.,** Role of blastocoel microenvironment in early mouse embryo differentiation, *Nature (London),* 284, 550, 1980.
34. **Potter, D. D., Furshpan, E. J., and Lennox, E.,** Connections between cells of the developing squid as revealed by electrophysiological methods, *Proc. Natl. Acad. Sci. U.S.A.,* 55, 328, 1966.
35. **Warner, A. E., Guthrei, S. C., and Gilula, N. B.,** Antibodies to gap-junctional protein selectively disrupt junctional communication in the early amphibian embryo, *Nature (London),* 311, 127, 1984.
36. **Hörstadius, S.,** The mechanics of sea urchin development, studied by operative methods, *Biol. Rev.,* 14, 132, 1939.
37. **Fink, R. D. and McClay, D. R.,** Three cell recognition changes accompany the ingression of sea urchin primary mesenchyme cells, *Dev. Biol.,* 107, 66, 1985.
38. **De Simone, D. W. and Spiegel, M.,** Micromere-specific cell surface proteins of 16-cell stage sea urchin embryos, *Exp. Cell Res.,* 156, 7, 1985.
39. **Karp, G. C. and Solursh, M.,** Acid mucopolysaccharide metabolism, the cell surface and primary mesenchyme activity in the sea urchin embryos, *Dev. Biol.,* 41, 110, 1974.
40. **Gustafson, T. and Wolpert, L.,** Cellular mechanisms in the morphogenesis of the sea urchin larva. Change in shape of cell sheets, *Exp. Cell. Res.,* 27, 269, 1962.
41. **Gustafson, T. and Wolpert, L.,** Cellular movement and contact in sea urchin morphogenesis, *Biol. Rev.,* 42, 442, 1967.
42. **Kawabe, T. K., Armstrong, P. B., and Pollock, E. G.,** An extracellular fibrillar matrix in gastrulating sea urchin embryos, *Dev. Biol.,* 85, 509, 1981.
43. **Vogt, W.,** Gestaltungsanalyse am Amphibienkeim mit örtlicher Veitalfärbung. Vorwort über Wege und Ziele. I. Methodik und Wirkungsweise der örtlichen Vitalfärbung mit Agar als Farbträger, *Wilhelm Roux Arch.,* 106, 542, 1925.
44. **Vogt, W.,** Gestaltungsanalyse am Amphibienkeim mit örtlicher Vitalfärbung. II. Gastrulation und Mesodermbildung bei Urodelen und Anuren, *Wilhelm Roux Arch.,* 120, 385, 1929.
45. **Keller, R. E.,** The cellular basis of epiboly: an SEM study of deep cell rearrangement during gastrulation in *Xenopus laevis, J. Embryol. Exp. Morphol.,* 60, 201, 1980.
46. **Cooke, J.,** Properties of primary organization field in the embryo of *Xenopus laevis.* IV. Pattern formation and regulation following early inhibition of mitosis, *J. Embryol. Exp. Morphol.,* 30, 49, 1973.
47. **Rhumbler, L.,** Zur Mechanik des Gastrulationsforgänges, insbesondere der Invagination. Eine Entwicklungsmechanische Studie, *Wilhelm Roux Arch.,* 14, 401, 1902.
48. **Holtfreter, J.,** A study of the mechanics of gastrulation. I., *J. Exp. Zool.,* 94, 261, 1943.
49. **Holtfreter, J.,** A study of the mechanics of gastrulation. II. *J. Exp. Zool.,* 95, 171, 1944.
50. **Keller, R. E.,** An experimental analysis of the role of bottle cells and the deep marginal zone in gastrulation of *Xenopus laevis, J. Exp. Zool.,* 216, 81, 1981.
51. **Le Blanc, J. and Brick, I.,** Morphologic aspects of adhesion and spreading behaviour of amphibian blastula and gastrula cells, *J. Embryol. Exp. Morphol.,* 61, 145, 1981.
52. **Brick, I., Schaeffer, B. E., Schaeffer, H. I., and Gannaro, J. F., Jr.,** Electrokinetic properties and morphologic characteristics of amphibian gastrula cells, *Ann. N.Y. Acad. Sci.,* 238, 399, 1974.
53. **Boucaut, J. C., Bernard, B., Aubery, M., Bourrillon, R., and Chouillon, C.,** Concanavalin A binding to amphibian embryo and effect on morphogenesis, *J. Embryol. Exp. Morphol.,* 51, 63, 1979.
54. **Nakatsuji, N., Gould, A. C., and Johnson, K. E.,** Movement and guidance of migrating mesodermal cells in *Ambystoma maculatum* gastrulae, *J. Cell Sci.,* 56, 207, 1982.
55. **Nakatsuji, N. and Johnson, K. E.,** Ectodermal fragments from normal frog gastrulae condition substrata to support normal and hybrid mesodermal cell migration *in vitro, J. Cell Sci.,* 68, 49, 1984.
56. **Thiery, J-P., Duband, J.-L., Tucker, G., Darbiere, T., and Boucaut, J.-C.,** The role of fibronectin in cell migration during early vertebrate embryogenesis, *Prog. Clin. Biol. Res.,* 151, 187, 1984.
57. **Boucaut, J-C., Darbiere, T., Boulekbache, H., and Thiery, J-P.,** Prevention of gastrulation but not neurulation by antibodies to fibronectin in amphibian embryos, *Nature (London),* 307, 364, 1984.

58. **Boucaut, J-C., Darbiere, T., Poole, T. J., Aoyama, H., Yamada, K. M., and Thiery, J-P.,** Biologically active synthetic peptides as probes of embryonic development: a competitive peptide inhibitor of fibronectin function inhibits gastrulation in amphibian embryos and neural crest cell migration in avian embryos, *J. Cell Biol.,* 99, 1822, 1984.

59. **Nakatsuji, N., Smolira, M. A., and Wylie, C. C.,** Fibronectin visualized by scanning electron microscopy immunochemistry on the substratum for cell migration in *Xenopus laevis* gastrulae, *Dev. Biol.,* 107, 264, 1985.

60. **Waddington, C. H.,** *The Epigenetics of Birds,* Cambridge University Press, London, 1952.

61. **Romanoff, A. L.,** *The Avian Embryo,* Macmillan, New York, 1960.

62. **Spratt, N. T., Jr. and Haas, H.,** Morphogenetic movements in the lower surface of the unincubated and early blastoderm, *J. Exp. Zool.,* 144, 139, 1960.

63. **Nicolet, G.,** La chronologie d'invagination chez le poulet: etude à l'aide de la thymidine tritiée, *Experientia,* 23, 576, 1967.

64. **Nicolet, G.,** Analyse autoradiographique de la localisation des différentes ébouches présomptive dans la ligne primitive de l'embryon de poulet, *J. Embryol. Exp. Morphol.,* 23, 79, 1970.

65. **Modak, S. P.,** Sur l'origine de l'hypoblaste chez les oiseaux, *Experientia,* 21, 273, 1965.

66. **Modak, S. P.,** Analyse experimentale de l'origine de l'endoblaste embryonnaire chez les oiseaux, *Rev. Suisse Zool.,* 73, 877, 1966.

67. **Fontaine, J. and LeDouarin, N. M.,** Analysis of endoderm formation in the avian blastoderm by the use of quail-chick chimaeras, *J. Embryol. Exp. Morphol.,* 41, 209, 1977.

68. **Sanders, E. J., Bellairs, R., and Portch, P. A.,** *In vivo* and *in vitro* studies on the hypoblast and definitive endoblast of avian embryos, *J. Embryol. Exp. Morphol.,* 46, 187, 1978.

69. **Bellairs, R., Ireland, G. W., Sanders, E. J., and Stern, C. D.,** The behaviour of embryonic chick and quail tissues in culture, *J. Embryol. Exp. Morphol.,* 61, 15, 1981.

70. **Spratt, N. T., Jr. and Haas, H.,** Germ layer formation and the role of the primitive streak. I. Basic architecture and morphogenetic tissue movements, *J. Exp. Zool.,* 158, 9, 1965.

71. **Bellairs, R.,** Gastrulation process in the chick embryo, in *Cell Behaviour,* Bellairs, R., Curtis, A. S. G., and Dunn, G., Eds., Cambridge University Press, London, 1982, 395.

72. **Sanders, E. J.,** Labelling basement membrane constituents in the living chick embryo during gastrulation, *J. Embryol. Exp. Morphol.,* 79, 113, 1984.

73. **Mitrani, E.,** Primitive streak-forming cells of chick invaginate through a basement membrane, *Wilhelm Roux Arch.,* 191, 320, 1982.

74. **Sanders, E. J.,** Ultrastructural immunochemical localization of fibronectin in the early chick embryo, *J. Embryol. Exp. Morphol.,* 71, 155, 1982.

75. **Duband, J. L. and Thiery, J-P.,** Appearance and distribution of fibronectin during chick embryo gastrulation and neurulation,, *Dev. Biol.,* 94, 337, 1982.

76. **Harrisson, F., Vanrolen, Ch., Foidart, J.-M., and Vakaet, L.,** Expression of different regional patterns of fibronectin immunoreactivity during mesoblast formation in the chick blastoderm, *Dev. Biol.,* 101, 373, 1984.

77. **Sternberger, L. A., Hardy, P. H., Cuculis, J. J., and Meyer, H. G.,** The unlabeled antibody enzyme method of immunocytochemistry. Preparation and properties of soluble antigen-antibody complex (horseradish peroxidase-anti-horseradish peroxidase) and its use in identification of spirochetes, *J. Histochem. Cytochem.,* 18, 315, 1970.

78. **Amoroso, E. C.,** Viviparity, in *Cellular and Molecular Aspects of Implantation,* Glasser, S. R. and Bullock, D. W., Eds., Plenum Press, New York, 1981, 3.

79. **Sherman, M. I., Sellens, M. H., Atienza-Samols, S. B., Pai, A. C., and Schindler, J.,** Relationship between the programs for implantation and trophoblast differentiation, in *Cellular and Molecular Aspects of Implantation,* Glasser, S. R. and Bullock, D. W., Eds., Plenum Press, New York, 1981, 75.

80. **Rosenfield, M. G. and Joshi, M. S.,** Role of a uterine endopeptidase in the implantation process in the rat, in *Cellular and Molecular Aspects of Implantation,* Glasser, S. R. and Bullock, D. W., Eds., Plenum Press, New York, 1981, 423.

81. **Morris, J. E., Potter, S. W., Rynd, L. S., and Buckley, P. M.,** Adhesion of mouse blastocysts to uterine epithelium in culture: a requirement for mutual surface interactions, *J. Exp. Zool.,* 225, 467, 1983.

82. **Enders, A. C., Schlafke, S., and Welsh, A. O.,** Trophoblastic and uterine epithelial surfaces at the time of blastocyst adhesion in the rat, *Am. J. Anat.,* 159, 59, 1980.

83. **Chavez, D. J. and Enders, A. C.,** Temporal changes in lectin binding of preimplantation mouse blastocysts, *Dev. Biol.,* 87, 267, 1981.

84. **Pinsker, M. C. and Mintz, B.,** Change in cell surface glycoproteins of mouse embryos before implantation, *Proc. Natl. Acad. Sci. U.S.A.,* 70, 1645, 1973.

85. **Morris, J. E. and Potter, S. W.,** A comparison of developmental changes in surface charge in mouse blastocysts and uterine epithelium using DEAE beads and dextran sulfate *in vitro, Dev. Biol.,* 103, 190, 1984.

86. **Finn, C. A. and Publicover, M.,** Cell proliferation and cell death in the endometrium, in *Cellular and Molecular Aspects of Implantation,* Glasser, S. R. and Bullock, D. W., Eds., Plenum Press, New York, 1981, 181.

87. **Kubo, H., Spindle, A., and Pedersen, R. A.,** Inhibition of mouse blastocyst attachment and growth by protease inhibitors, *J. Exp. Zool.,* 216, 445, 1981.

88. **Axelrod, H.,** Altered trophoblast functions in implantation-defective mouse embryos, *Dev. Biol.,* 108, 185, 1985.

89. **Mossman, H. W.,** Orientation and site of attachment of the blastocyst: a comparative study, in *Biology of the Blastocyst,* Blandau, R. J., Ed., University of Chicago Press, 1971, 49.

90. **Smith, L. J.,** Embryonic axis orientation in the mouse and its correlation with blastocyst relationships to the uterus, *J. Embryol. Exp. Morphol.,* 55, 257, 1980.

91. **Rawles, M.,** Tissue interactions in scale and feather development as studied in dermal epidermal recombinations, *J. Embryol. Exp. Morphol.,* 11, 765, 1963.

92. **Wessels, N. K.,** Mammalian lung development: interactions in formation and morphogenesis of tracheal buds, *J. Exp. Zool.,* 175, 155, 1970.

93. **Spooner, B. S.,** Mammalian lung development: interactions in primordium formation and bronchial morphogenesis, *J. Exp. Zool.,* 175, 445, 1970.

94. **Decker, R. S.,** Disassembly of zonula occludens during amphibian neurulation, *Dev. Biol.,* 81, 12, 1981.

95. **Gläser, O. C.,** On the mechanism of morphological differentiation in nervous system, *Anat. Rec.,* 8, 525, 1914.

96. **Burnside, B.,** Microtubules and microfilaments in amphibian neurulation, *Am. Zool.,* 13, 989, 1973.

97. **Schroeder, T. E.,** Cell constriction: contractile role of microfilaments in division and development, *Am. Zool.,* 13, 949, 1973.

98. **Schoenwolf, G. C. and Franks, M. V.,** Quantitative analysis of changes in cell shapes during bending of the avian neural plate, *Dev. Biol.,* 105, 257, 1984.

99. **Mulherkar, L., Rao, K. V., and Joshi, S. S.,** Studies on some aspects of the role of sulfhydryl groups in morphogenesis, *J. Embryol. Exp. Morphol.,* 14, 129, 1965.

100. **Mulherkar, L. and Rao, K. V.,** The role of thiols in the primary embryonic organizer action in vertebrate embryos, *J. Sci. Ind. Res.,* 34, 511, 1975.

101. **Jacobson, C-O.,** Experiments on β-mercaptoethanol as an inhibitor of neurulation movements in the amphibian neurula, *J. Embryol. Exp. Morphol.,* 23, 463, 1970.

102. **Jacobson, A. G. and Gordon, R.,** Changes in the shape of the developing vertebrate nervous system analyzed experimentally, mathematically and by computer simulation, *J. Exp. Zool.,* 197, 191, 1976.

103. **Odell, G. M., Oster, G., Alberch, P., and Burnside, B.,** The mechanical basis of morphogenesis. I. Epithelial folding and invagination, *Dev. Biol.,* 85, 446, 1981.

104. **Sauer, F. C.,** Interkinetic migration of embryonic epithelial nuclei, *J. Morphol.,* 60, 1, 1936.

105. **Brown, M. G., Hamburger, V., and Schmidt, F. O.,** Density studies on amphibian embryos with special reference to the mechanism of organizer action, *J. Exp. Zool.,* 88, 353, 1941.

106. **Childress, S.,** Models of cell interaction based on differential adhesion, *J. Biochem. Eng.,* 106, 36, 1984.

107. **Bellairs, R., Curtis, A. S. G., and Sanders, E. J.,** Cell adhesiveness and embryonic differentiation, *J. Embryol. Exp. Morphol.,* 46, 207, 1978.

108. **Bahl, K. N.,** Excretion in Oligochaeta, *Biol. Rev.,* 22, 109, 1947.

109. **Mendrick, D. L., Rennke, H. G., Cortan, R. S., Springer, T. A., and Abbas, A. K.,** Monoclonal antibodies against rat glomerular antigens: production and specificity, *Lab. Invest.,* 49, 107, 1983.

110. **Kerjaschki, D., Sharkey, D. J., and Farquhar, M. G.,** Identification and characterization of podocalyxin — the major sialoprotein of the renal glomerular epithelial cells, *J. Cell Biol.,* 98, 1591, 1984.

111. **Blau, E. B. and Haas, J. E.,** Glomerular sialic acid and proteinuria in human renal disease, *Lab. Invest.,* 28, 477, 1973.

112. **Seiler, M. W., Rennke, H. G., Venkatachalam, M. A., and Cotran, R. S.,** Pathogenesis of polyanion-induced alterations ("fusion") of glomerular epithelium, *Lab. Invest.,* 36, 48, 1977.

113. **Goldin, G. V., Hindman, H. M., and Wessels, N. K.,** The role of cell proliferation and cellular shape change in branching morphogenesis of the embryonic mouse lung: analysis using aphidicolin and cytochalasins, *J. Exp. Zool.,* 232, 287, 1984.

114. **Goldin, G. V.,** Towards a mechanism for morphogenesis in epithelio-mesenchymal organs, *Q. Rev. Biol.,* 55, 251, 1980.

115. **Bernfield, M. R. and Benerjee, S. D.,** The basal lamina in epithelial-mesenchymal morphogenetic interactions, in *Biology and Chemistry of Basement Membranes,* Kaffalides, N. A., Ed., Academic Press, New York, 1978, 137.

116. **Cohn, R. H., Banerjee, S. D., and Bernfield, M. R.,** Basal lamina of salivary epithelia. Nature of glycosaminoglycan and organization of extracellular materials, *J. Cell Biol.,* 73, 464, 1977.

117. **Gordon, J. R. and Bernfield, M. R.,** The basal lamina of the postnatal mammary epithelium contains glycosaminoglycans in a precise ultrastructural organization, *Dev. Biol.,* 74, 118, 1980.

118. **Thomson, H. A. and Spooner, B. S.,** Inhibition of branching morphogenesis by alteration of glycosaminoglycan biosynthesis in salivary glands treated with β-D-xyloside, *Dev. Biol.,* 89, 417, 1982.

119. **Wicha, M. S., Liotta, L. A., Vonderhaar, B. K., and Kidwell, W. R.,** Effect of inhibition of basement membrane collagen deposition on rat mammary gland development, *Dev. Biol.,* 80, 253, 1980.

120. **Spooner, B. S. and Faubion, J. M.,** Collagen involvement in branching morphogenesis of embryonic lung and salivary gland, *Dev. Biol.,* 77, 84, 1980.

121. **Toole, B. P.,** Glycosaminoglycans in morphogenesis, in *Cell Biology of Extracellular Matrix,* Hay, E. D., Ed., Plenum Press, New York, 1981, chap. 9.

122. **Menkes, B. and Sandor, S.,** Somitogenesis: regulation potencies, sequence determination and primordial interactions, in *Vertebrate Limb and Somite Morphogenesis,* Ede, D. A., Hinchliffe, J. R., and Balls, M., Eds., Cambridge University Press, London, 1977, 403.

123. **Flint, O. P., Ede, D. A., Wilby, O. K., and Proctor, J.,** Control of somite number in normal and amputated embryos: an experimental and theoretical analysis, *J. Embryol. Exp. Morphol.,* 45, 189, 1978.

124. **Veini, M. and Bellairs, R.,** Experimental analysis of control mechanisms in somite segmentation in avian embryos. I. Reduction of material at the blastula stage in *Coturnix coturnix japonica, J. Embryol. Exp. Morphol.,* 74, 1, 1983.

125. **Bellairs, R. and Veini, M.,** An experimental analysis of somite segmentation in the chick embryo, *J. Embryol. Exp. Morphol.,* 55, 93, 1980.

126. **Bellairs, R. and Veini, M.,** Experimental analysis of control mechanisms in somite segmentation in avian embryos. II. Reduction of material in the gastrula stages of the chick, *J. Embryol. Exp. Morphol.,* 79, 183, 1984.

127. **Cooke, J. and Zeeman, E. C.,** A clock and wave front model for the control of the number of repeated structures during animal morphogenesis, *J. Theoret. Biol.,* 58, 455, 1976.

128. **Cooke, J.,** The control of somite number during amphibian development: models and experiments, in *Vertebrate Limb and Somite Morphogenesis,* Ede, D. A., Hinchliffe, J. R., and Balls, M., Eds., Cambridge University Press, London, 1977, 433.

129. **Youn, B. W. and Malacinski, G. M.,** Somitogenesis in the amphibian *Xenopus laevis.* Scanning electron microscopic analysis of intrasomitic cellular arrangements during somite rotation, *J. Embryol. Exp. Morphol.,* 64, 23, 1981.

130. **Hamilton, L.,** The formation of somites in *Xenopus, J. Embryol. Exp. Morphol.,* 22, 253, 1969.

131. **Bellairs, R.,** The mechanism of somite segmentation in the chick embryo, *J. Embryol. Exp. Morphol.,* 51, 227, 1979.

132. **Elsdale, T., Pearson, M., and Whitehead, M.,** Abnormalities in somite segmentation following heat shock to *Xenopus* embryos, *J. Embryol. Exp. Morphol.,* 35, 625, 1976.

133. **Cheney, C. M. and Lash, J. W.,** An increase in cell-cell adhesion in the chick segmental plate results in meristic pattern, *J. Embryol. Exp. Morphol.,* 79, 1, 1984.

134. **Lash, J. W., Seitz, A. W., Cheney, C. M., and Ostrovsky, D.,** On the role of fibronectin during the compaction of somitogenesis in the chick embryo, *J. Exp. Zool.,* 232, 197, 1984.

135. **Kvist, T. N. and Finnegan, C. V.,** The distribution of glycosaminoglycans in the axial region of the developing chick embryo. I. Histochemical analysis, *J. Exp. Zool.,* 175, 221, 1970.

136. **Kvist, T. N. and Finnegan, C. V.,** The distribution of glycosaminoglycans in the axial region of the developing chick embryo. II. Biochemical analysis, *J. Exp. Zool.,* 175, 241, 1970.

137. **Hay, E. D.,** Fine structure of embryonic matrices and their relation to the cell surface in ruthenium red-fixed tissues, *Growth,* 42, 399, 1978.

138. **Hay, E. D.,** Collagen and embryonic development, in *Cell Biology of Extracellular Matrix,* Hay, E. D., Ed., Plenum Press, New York, 1981, chap. 12.

139. **Solursh, M., Fisher, M., Meier, S., and Singley, C. T.,** The role of extracellular matrix in the formation of the sclerotome, *J. Embryol. Exp. Morphol.,* 54, 75, 1979.

140. **Greenberg, G. and Hay, E. D.,** Epithelia suspended in collagen gels can lose polarity and express characteristics of migrating mesenchymal cells, *J. Cell Biol.,* 95, 333, 1982.

141. **Braver, N. B.,** *The Mutants of Drosophila melanogaster Classified According to Body Parts Affected,* Publ. No. 552A, Carnegie Institute of Washington, Washington, D.C., 1956.

142. **Du Porte, E. M.,** The comparative morphology of the insect head, *Annu. Rev. Entomol.,* 2, 55, 1958.

143. **Anderson, D. T.,** The development of holometabolous insects, in *Developmental Systems: Insects,* Vol. 1, Counce, S. J. and Waddington, C. H., Eds., Academic Press, London, 1972, chap. 4.

144. **Jura, C.,** Development of apterygote insects, in *Developmental Systems: Insects,* Vol. 1, Counce, S. J. and Waddington, C. H., Eds., Academic Press, London, 1972, chap. 2.

145. **Lewis, E. B.,** Control of body segment differentiation in *Drosophila* by the bithorax gene complex, in *Embryonic Development: Genes and Cells,* Burger, M., Ed., Alan R. Liss, New York, 1982, 269.

146. **Hafen, E., Levine, M., and Gehring, W. J.,** Regulation of *antennapedia* transcript distribution by the *bithorax* complex in *Drosophila, Nature (London),* 307, 287, 1984.

147. **Kaufman, T. C. and Abbot, M. K.,** Homeotic genes and the specification of segmental identity in the embryo and adult thorax of *Drosophila melanogaster,* in *Molecular Aspects of Early Development,* Malacinski, G. M. and Klein, W. H., Eds., Plenum Press, New York, 1984, 189.

Chapter 9

NEOPLASIA

I. INTRODUCTION

"Nothing is more beautiful than an embryo developing in its harmonious way; nothing is uglier than a cancer growing in its malignant way."[1] Yet, ample evidence exists to show that there are many characteristics common to neoplastic cells and embryonic cells. How malignant tumor cells differ from normal adult cells and embryonic cells is therefore a fundamental question which has been approached by several generations of cancer researchers. If we know the basic mechanism by which a mass of cells builds itself into something as beautiful as an embryo, we can possibly point out the unique characteristics of the tumor cells that make them malignant. This is the hope which has sustained considerable part of cancer research. The notion that cancer is a disease characterized by disorganized embryonic differentiation seems to underlie most of the current discussions on the etiology of the disease.

In order to define the various aspects of the present discussion in relation to the cell surface and neoplastic change, a framework will have to be presented wherein we identify a number of discrete changes from the initiation of the disease to its most dangerous state, manifested by metastatic spread and the establishment of secondary growth centers. In view of the variety of tissues from which tumors can arise and the different degrees of deviation from normal cells observed in human and animal cancers, we shall have to consider a very generalized sequence of changes. Figure 1 presents such a sequence through which a normal tissue transforms into a cancerous one and establishes new centers of growth by metastatic spread.

II. ETIOLOGY

A priori, several causative factors leading to the transformation of normal cells into cancerous ones can be suggested. These include (1) constitutional lesions, (2) somatic mutations, (3) viral transformation of the genome, and (4) action of chemical carcinogens. From time to time these suggestions have found adequate support. Consequently they have served as working hypotheses for research on the problem. It must be pointed out, however, that the vast variety of human and animal tumors offer "typical" examples for all the hypotheses. It is therefore judicious to consider that the different hypotheses constitute emphases on different aspects of the etiology of cancers.

A. Constitutional Lesions

Many cancerous growths, which have no defined external causal factor, can be considered as arising from constitutional lesions. Experimental animals, especially certain inbred strains of mice, are known to develop tumors "spontaneously", and it is possible to argue that there is some constitutional deficiency in these animals. A general hypothesis can be proposed that any organism with a deficient immune system would fail to eliminate the body cells of a transformed phenotype and hence would be prone to develop "spontaneous" tumors which in normal individuals would be destroyed before they have grown to a detectable size. Though cancer is not a typical hereditary disease, proneness to certain types of cancers in highly inbred ethnic groups and experimental animals is an observed fact. This could be due to some constitutional defect. High susceptibility to carcinogenic mutagens is also a constitutional deficiency.

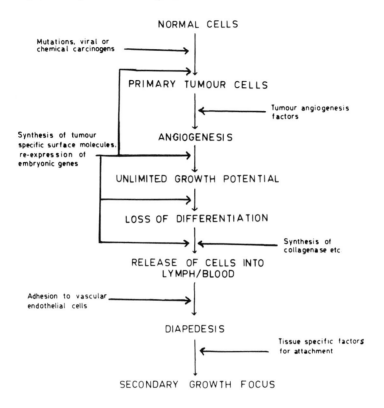

FIGURE 1.   Diagram depicting the sequence of some of the changes that lead to the transformation of normal cells into malignant cells. The vertical arrows in the middle represent a possible sequence and does not imply that all cancers invariably pass through it. See the text.

B. Somatic Mutations and Mutagenic Agents

By the time they are clinically detectable, most cancers appear monoclonal, i.e., they appear to be clones of individual progenitor cells. This would immediately suggest that the basic lesion causing neoplastic growth may be a mutation. The famous Ames test[2] is an important methodology to evaluate the mutagenic potential of suspected carcinogens and other factors. The genetic lesion may be a subtle point mutation or may be an overt change in the chromosomes. Different karyotypic abnormalities are associated with certain cancers, and this supports the idea that the initial cause is a genetic lesion. Not all cancers are, however, associated with a mutation. Highly malignant teratocarcinoma cells can participate in perfectly normal embryonic development, as shown by Mintz and associates, in allophenic mice.[3] In the mosaic individuals, the cells derived from a teratocarcinoma, which had gone through more than 200 transplant generations spanning 8 years, have been shown to differentiate normally and give rise to a variety of tissue cells.[4,5]

C. Viral Agents

Both DNA and RNA viruses are known to infect cells and transform them. The viruses are called oncogenic. The Epstein-Barr virus is a well-known example of DNA viruses causing human infectious mononucleosis.[6,7] Oncogenic RNA viruses (oncorna viruses) are associated with a number of animal neoplasms. The best documented example of such a virus is the Rous sarcoma virus (RSV). Chickens and several other avian species are susceptible to this viral infection. A unique variety of enzymes carried by the viruses provide for the synthesis of a double stranded DNA transcript of the viral genome in the infected host cell. The DNA copy is called the provirus. It becomes covalently linked and thus integrated into the host genome. The infected cells, which are now transformed, produce more virus RNA copies. A coat material is formed around the virus particles and the newly formed viruses are budded off from the cell surface. Other examples of oncorna viruses include the Friend focus virus, causing erythro- and myelo-monocytic tumors, and the Abelson leukemia virus, causing hematopoietic tumors in rodents. A variety of T-cell leukemias in rodents are caused by RNA viruses.[8-10]

D. Chemical Carcinogens

It has been known for a long time that certain cancers are caused by external agents. Nasal polyps and cancers are associated with the use of tobacco snuff and lung cancers with smoking. A large variety of skin cancers are associated with exposure to various polycyclic hydrocarbons which occur in coal tar.[11] A large number of chemicals in our environment have now been proved to be carcinogens or potential carcinogens. A variety of ''life style'' cancers may be traced to dietary factors.[12] There are many chemicals that do not cause cancerous lesions by themselves. Yet, if the exposed parts are subsequently treated with some other chemicals, called promoters, cancers develop in the affected tissue. From this a general concept of two-stage carcinogenesis has developed. According to this, a carcinogen, even by a single application, can change the tissue in a subtle manner potentiating it to respond to another chemical agent, viz., the promoter. The initial action of the carcinogen is generally irreversible. In the absence of a promoter, the carcinogen does not cause any visible lesion. Promotion is generally a prolonged process, i.e., several repeated exposures are required to bring about the disease. However, the promoter by itself is not a carcinogen.

The availability of chemicals that could induce the cancer of various target tissues in experimental animals and cells in vitro has greatly aided in elucidating their action.[13] ''Carcinogenic'' chemicals which affect the genome directly or through their metabolized derivatives are, strictly speaking, mutagens rather than carcinogens.

III. TUMOR ANGIOGENESIS

Whatever be the primary cause of neoplastic transformation of cells, the development of all the pathological conditions associated with the disease are regulated by several other factors. When a cell escapes from its normal growth control and continues to grow by repeated mitotic divisions, the primary neoplastic focus may be considered to have been established. Abundance of nutrients is an essential condition for cell multiplication. The progenitor cell presumably has an advantage over the other healthy cells in acquiring and utilizing the available nutrients.[14] As a result, its clone will grow faster and soon become a small spheroid of cells. Despite the acquired ability to grow fast, a limit is eventually imposed on the growth of the spheroid since the nutrients from the surrounding tissue fluids cannot diffuse and reach the individual cells to meet their requirement easily. It must be noted that while the surface of the spherical cell mass increases by the second power, the mass increases by the third power. Consequently the growth of the primary tumor cells is a self-limiting process. Further growth of the cells depends on the development of an adequate vascular supply penetrating the cellular mass. Angiogenesis, i.e., the formation of vascular structures, is thus a very important factor that contributes to the growth of solid tumors.[15] In case of transformed cells which are free, i.e., not as a solid mass, dissemination can start immediately.

The sequence of changes leading to the neovascularization will be presented very briefly. It was known for a long time that new capillaries arise in association with tumors. However, the immediate cause of this event was identified only recently. Folkman[16] demonstrated that tumors implanted into chicken embryos induce angiogenesis in the host tissue. It was also reported by several workers that fast-dividing embryonic cells can cause neovascularization when implanted in a host tissue.[17] From this it was suggested that tumor cells produce an angiogenesis factor. The cells responding to this factor were identified as the capillary endothelial cells in the vicinity of the tumor. In this context, it must be pointed out that the capillary endothelial cells are structurally and functionally different from aortic endothelial cells.[18,19] The tumor angiogenesis factors seem to act specifically on the capillary endothelium and promote their penetration into the mass of the tumor.

Considerable work has been done on the origin and mode of action of the angiogenetic factor(s). Though it is generally held that the factor is produced by the tumor cells, the possibility of its production by other cells has also been demonstrated.[20] Mast cells are found to accumulate at the tumor site before the ingrowth of new capillaries.[21] The mast cells release heparin, which has been found to enhance the migration of capillary endothelial cells in vitro[19] and neovascularization in the chorioallantoic membrane of the chick embryo.[22,23] Protamine, an arginine-rich basic protein, is able to bind heparin and inhibit angiogenesis in vivo. The heparin-stimulated migration of capillary endothelial cells in vitro is also inhibited by protamine.[22] Heparin has affinity for the angiogenesis factor. It is probable that this affinity is crucial to neovascularization. The heparin affinity has been used in a technique of heparin-affinity chromatography for the purification of the factor.[24,25] Most solid tumors are angiogenesis-dependent, and hence the mechanism of this process has been studied with great zeal with the hope that if one can inhibit the process, it could constitute an effective measure against the growth of a primary tumor. Folkman et al.[23] have shown that heparin combined with cortisone can inhibit tumor angiogenesis. Even oral administration of these drugs seems to be effective, since the inhibitory effect is due to a heparin fragment and not the large heparin molecule. The precise mode of action of this combination of the drugs is, however, not clear. Further, it is not yet at a stage of clinical application.

IV. CELL SURFACE CHANGES AND METASTASIS

A localized tumor can often be removed surgically. However, if the tumor spreads to distant anatomical sites and establishes new centers of growth, a surgical cure is virtually impossible. It is the dissemination or metastasis of the tumors that is the ugliest of their attributes. A clear understanding of how such a property is acquired by the tumor cells is therefore the ultimate goal of fundamental research on cancer. In the previous chapters, we have reviewed the evidence indicating that embryonic cells possess specific mechanisms whereby they can detach from an anatomical site, migrate over long distances, and occupy new locations. The avian primordial germ cells take a vascular route and eventually get lodged in the gonad area. There are other examples of cells making their way through other tissues. These developmental activities can be analyzed fruitfully as models of metastatic spread. Spreading by permeating into neighboring tissues is found in many cancers. In fact the word *cancer* is derived from this feature, viz., a tumor spreading its arms in all directions like a crab. There are also examples of tumor cells taking a lymphatic or blood vascular route. Investigating the adhesive and motile properties of the cells is obviously important. Relatively stable cells of a primary tumor may at a later stage acquire changes in their adhesive and motile behavior, leading to metastasis. Thus a limited similarity exists between some embryonic cells and tumor cells. A comparison of the adhesive and motile behavior of embryonic cells with tumor cells therefore assumes immediate relevance. There are many instances of tumor cells acquiring some attributes of embryonic cells. That is why cancer is considered an example of retrodifferentiation. An interesting analysis of this notion has been discussed elaborately by Sherbet.[26]

A. Synthesis of Tumor-Specific Surface Molecules

A search for tumor-specific characteristics, especially at the cell surface, is obviously a promising approach to an analysis of metastasis. Important differences have been reported between malignant cells and their healthy counterparts.[27] Fetal antigens (alpha fetoprotein, which is characteristic of hepatomas, testicular, and ovarian tumors, etc.; carcinoembryonic antigen, characteristic of the carcinomas of pancreas, colon, and rectum), ectopic hormones, a variety of glycosylating enzymes, and glycolytic enzymes are associated with tumors and other non-neoplastic pathological conditions.[26]

However, which of these differences are causally related to malignancy is a moot point. In some cases there is excellent evidence to show a causal relationship rather than mere correlation. For example, a 37,000 dalton plasma membrane glycoprotein from mouse sarcoma cells has been shown to be causally related to the invasive behavior of these cells. Their invasive behavior can be prevented by Fab fragments of antibodies against the surface glycoprotein.[28] In general, surface glycoproteins, which can alter the biophysical properties or specific recognition sites, are likely to be of special relevance to the question. Yogeeswaran and Salk[29] have observed higher sialic acid content in malignant cells compared with non-malignant ones. The sialic acid exposed to the cell surface increases significantly in the cells of higher metastatic potential. In particular, sialylation of the cell surface galactosyl and N-acetyl galactosaminosyl residues increases. Yogeeswaran and Salk[29] have suggested several possible mechanisms through which the increased sialylation could contribute to the metastatic behavior. Since cell surface glycoconjugates are undoubtedly involved in a variety of cellular interactions such as specific adhesion and recognition of their cellular and extracellular milieu, the discovery of any difference in their content, distribution, and synthesis between malignant and nonmalignant cell types is of interest. An increase in the sialic acid residues on the surface carbohydrates of cells will increase the negative charge density and is hence expected to increase their electrophoretic mobility or decrease their pI. Sherbet,[26] who has discussed some of the investigations to verify this proposition, has concluded that in metastasizing lymphosarcoma there is an increase in the negative charge density compared with its nonmetastasizing counterpart. However, this cannot be considered as a generalization applicable to all tumors.[26]

B. Disruption of Cell-Cell Contacts

Cells positioned in a tissue in close association with other cells and the extracellular matrix will not move away from their location unless loosened in some manner. Gap junctions, which establish intimate intercellular communication, and desmosomes, which provide for close and stabilized association of neighboring plasma membranes, could resist such loosening. Besides, specific covalent and noncovalent binding of the molecules of the extracellular matrix among themselves and with the cell surface would tend to stabilize the position of cells. Disappearance of gap junctions along with desmosomal complexes in response to the application of a potent tumor promoter in a chemically induced mouse cell carcinoma is an interesting observation in this context.[30] Similar effects have been demonstrated in vitro by other workers.[31,32]

C. Destabilization of Epithelia

The association of cells among themselves and with the extracellular material is important in morphogenesis. Stability of such associations is undoubtedly the basis of tissue architecture attained after differentiation. Epithelia constitute only a small fraction of the total body weight. Yet, carcinomas (cancers arising from epithelia) comprise more than 70% of clinically manifest cancers. Continued growth through mitosis, as well as destabilization of the cells preliminary to metastasis, would depend on specific perturbation of the epithelial tissues. The epithelial character of the cells is maintained by their mutual adhesion through specialized junctional complexes and the basement membranes consisting of Type IV collagen, laminin, fibronectin, etc. Destabilization of the basement membrane is essential for the epithelial cells to loosen themselves and enter into the neighboring tissue. Tomasek et al.[33] observed that a number of embryonic epithelia can migrate into a collagen gel in which they are placed. This is achieved by holding the epithelial cells in a gelling solution of collagen. In this condition, the cells transdifferentiate into mesenchyme. It has been suggested that the apical surface of the epithelial cell reacts with the collagen fibers, leading to a change in the membrane characteristics.

Information on how the epithelial cells maintain their cell-cell and cell-basement membrane contacts will have to be obtained for a clearer understanding of the mechanism that causes the destabilization of the epithelial morphology. A recent report[34] on experimental manipulation of the morphology of an established cell line, MDCK, of canine kidney epithelial origin, may be mentioned as an interesting achievement in this context. Using a monoclonal antibody (Anti-Arc 1), Imhoff et al.[34] have shown that it is possible to perturb the mutual contacts of the MDCK cells. Treatment with the antibody results in loosening of their junctions and loss of polarity. Thus the antibody identifies a cell surface adhesive mechanism that holds the epithelial cells *in situ*. Destabilization of epithelial cells may therefore be attributed to a defect in their adhesive mechanism.

Dissemination of cells by passing through capillary walls would also depend on the destabilization of basement membranes. It has been demonstrated that malignant cells are able to traverse the basement membrane.[35,36] The expression of transformation-associated proteases, which degrade fibronectin at cell contact sites, has been reported.[37] Fragmentation of fibronectin can lead to loosening of the cell's hold on collagen. Besides, the fibronectin fragments may stimulate mitotic activity in the cells.[38] A somewhat analogous situation has been discussed in the previous chapter in connection with the branching morphogenesis of tubular epithelia. It seems that the complex regulatory role of the extracellular matrix on the morphogenetic activities of cells is just beginning to be visualized, and more needs to be learned about it. In particular, the basic mechanisms, which destabilize epithelial cells such as those in the neural fold of vertebrate embryos[39] and the primitive streak of the blastoderm,[40,41] ought to be analyzed further. An attractive aspect of the neural crest development is that there are several features of similarity between it and the development of carcinomas.

D. Loosening of Tumor Cells by Proteolysis

A limited proteolytic activity in the vicinity of the primary tumor can facilitate the dissemination of its cells. This notion finds acceptance in many discussions on the mechanisms of metastasis. A short treatment with proteolytic enzymes is known to release cell monolayers in vitro from contact inhibition of growth and thereby facilitate an increase in their saturation densities. Different proteolytic enzyme activities have been detected in a variety of tumors and transformed cells in vitro (Table 1). In particular, the presence of collagenase and plasminogen activator in tumors is of interest. We may recall from the previous chapter that plasminogen activator is found associated with the mouse blastocyst also. Possibly it is a general mechanism of implantation. The invasive ability of the trophoblast cells as well as of the tumor cells may therefore be attributed to the proteolytic enzyme activities associated with them. Thus the view that proteolytic activity may be an important factor in tissue invasiveness, in general, has gained ground. Consistent with this is the finding of Latner et al.[49,50] that aprotinin, a general inhibitor of proteolytic activity, can afford protection against metastatic spread of cancers.

Cells may be loosened from their location in the tissues by a decrease in their adhesiveness and/or increased motility. In fact, the notion that invasive cancerous cells are characterized by decreased mutual adhesiveness has gained general acceptance. Coman[51] was one of the earliest to suggest this by demonstrating that the cells of squamous carcinoma have greatly decreased adhesiveness. Subsequently many reports appeared in the literature indicating that

Table 1
SUGGESTED ROLE FOR PROTEOLYTIC ENZYMES IN TUMOR METASTASIS

Tumor/transformed cells	Enzyme(s)	Ref.
3T3 Py virus-transformed fibroblasts	Proteolytic enzymes	42
DNA/RNA virus-transformed fibroblasts	Fibrinolytic enzymes	43, 44
Rous Sarcoma virus-transformed chick fibroblasts	Plasminogen activator	45
Dimethylbenzanthracene-caused rat mammary carcinoma	Plasminogen activator	46
Ascites hepatomas	Neutral proteases	47
Malignant melanoma	Collagenase	48

a decreased cell adhesiveness may be correlated with invasiveness. In Chapter 2 we discussed the phenomenon of cell adhesion. It may be recalled that several distinct cell surface proteins/glycoproteins are involved in the establishment of cell contacts. These in turn may be stabilized in the plasma membrane by the anchoring of cytoskeletal elements. Any alteration in this machinery may lead to defective adhesiveness and/or increased cell motility. It has been shown that phosphorylation of the tyrosine residues of vinculin in Rous Sarcoma virus-transformed cells leads to a decrease in its effectiveness as a linker molecule at the adhesion plaques. Presumably this leads to an alteration of the adhesive properties and general morphology of the transformed cell.[52] Defect in adhesion plaques is probably one of several other mechanisms by which cell adhesiveness is modified during transformation.[53,54]

Once the cells are loosened from their structural associations in the tissue, they can express their motile behavior. This may lead them into the lymphatics or blood vessels. Since the lymphatic and blood circulatory systems are interlinked, the result will be the eventual dispersal of the cells in the entire body. Clinical and experimental studies have indicated that the secondary foci of growth may be established in remote parts of the body, much away from the location of the primary tumor. However, there are many examples of tumors metastasizing preferentially to one or more specific organs.[55] Besides routine histopathological demonstration of the secondary growth centers in the target tissues, recently developed immunocytochemical methods can aid the detection of even single tumor cells. For example, the "epithelial membrane antigen" is present in a variety of normal human epithelial tissues and many carcinomas. Specific fluorescent antibodies raised against the antigen can identify normal and neoplastic epithelia in cryostat or paraffin embedded tissue sections. Experimentally, a suspension of the tumor cells may be placed over fully confluent monolayers of different test cell types to detect the preferential adhesion of cells. Organ-specific binding of tumor cells has been demonstrated even in cryostat sections.[56] Obviously organ-specific lodgement of the tumor cells depends on the nature of the peculiar environment offered by the target tissues or organs.

V. CONCLUDING REMARKS

The brief discussion presented in this chapter is adequate to show that carcinogenesis consists of a multistep morbid differentiation of normal cells through various grades of malignancy. The cell surface plays a key role in all the steps. Embryonic development provides many aspects that are analogous, albeit to a limited degree, to the process of tumor development. Identification and analysis of the discrete developmental process associated with neoplastic growth and dissemination can offer clues for intervention. Emphasizing these points was the primary objective of including this chapter in the present book.

REFERENCES

1. **Brachet, J.,** *Biochemical Cytology,* Academic Press, New York, 1957, 446.
2. **Ames, B. N., Durston, W. E., Yamasak, E., and Lee, F. D.,** Carcinogens are mutagens: a simple test system combining liver homogenates for activation and bacteria for detection, *Proc. Natl. Acad. Sci. U.S.A.,* 70, 2281, 1973.
3. **Mintz, B.,** Genetic mosaicism and *in vivo* analysis of neoplasia and differentiation, in *Cell Differentiation and Neoplasia,* Saunders, G. F., Ed., Raven Press, New York, 1978, 27.
4. **Mintz, B. and Illmensee, K.,** Normal genetically mosaic mice produced from malignant teratocarcinoma cells, *Proc. Natl. Acad. Sci. U.S.A.,* 72, 3585, 1975.
5. **Illumensee, K. and Mintz, B.,** Totipotency and normal differentiation of single teratocarcinoma cells cloned by injection into blastocysts, *Proc. Natl. Acad. Sci. U.S.A.,* 73, 549, 1976.
6. **Epstein, M. A.,** Epstein-Barr virus as the cause of human cancer, *Nature (London),* 274, 740, 1978.
7. **De Thé, G., Gesser, A., Day, N. E., Tukei, P. M., Williams, E. H., Beri, D. P., Smith, P. G., Dean, A. G., Bornkamm, G. W., Feorino, P., and Henle, W.,** Epidemiological evidence for causal relationship between Epstein-Barr virus and Burkitt's lymphoma from Ugandan prospective study, *Nature (London),* 274, 756, 1978.
8. **Dulbecco, R.,** From the molecular biology of oncogenic DNA viruses to cancer, *Science,* 192, 437, 1976.
9. **Baltimore, D.,** Viruses, polymerases and cancer, *Science,* 192, 632, 1976.
10. **Temin, H. M.,** The DNA provirus hypothesis, *Science,* 192, 1075, 1976.
11. **Yuspa, S. H., Hennings, H., and Saffiotti, U.,** Cutaneous chemical carcinogenesis: past, present and future, *J. Invest. Dermatol.,* 67, 197, 1976.
12. **Grahams, S., Dayal, H., Swanson, M., Mittelman, A., and Wilkinson, G.,** Diet and epidemiology of the cancer of colon and rectum, *J. Natl. Cancer Inst.,* 51, 709, 1978.
13. **Miller, E. C. and Miller, J. A.,** Searches for ultimate chemical carcinogens and their reaction with cellular macromolecules, *Cancer,* 47, 2327, 1981.
14. **Bhargava, P. M.,** Regulation of cell division and malignant transformation: a new model for control by uptake of nutrients, *J. Theoret. Biol.,* 68, 101, 1977.
15. **Gimbrone, M. A., Jr., Leapman, S. B., Cotran, R. S., and Folkman, J.,** Tumor dormancy *in vivo* by prevention of neovascularization, *J. Exp. Med.,* 136, 261, 1972.
16. **Folkman, J.,** Tumor angiogenesis factor, *Cancer Res.,* 34, 2109, 1974.
17. **Keegan, A., Hill, C., Kumar, S., Phillips, P., Schor, A., and Weiss, J.,** Purified angiogenesis factor enhances proliferation of capillary, but not aortic endothelial cells *in vitro, J. Cell Sci.,* 55, 261, 1982.
18. **Zetter, B. R.,** Migration of capillary endothelial cells is stimulated by tumour-derived factors, *Nature (London),* 285, 41, 1980.
19. **Folkman, J. and Haudenschild, C.,** Angiogenesis *in vitro, Nature (London),* 288, 551, 1980.
20. **Polverini, P. J., Cotran, R. S., Gimbrone, M. A., Jr., and Unanue, E. R.,** Activated macrophages induce vascular proliferation, *Nature (London),* 269, 804, 1977.
21. **Fraser, R. A. and Simpson, J. G.,** Role of mast cells in experimental tumour angiogenesis, *Ciba Foundation Symposium,* Symp. 100, Pitman Books, London, 1983, 120.
22. **Folkman, J., Taylor, S., and Spillberg, C.,** The role of heparin in angiogenesis, *Ciba Foundation Symposium,* 100, Pitman Books, London, 1983, 132.
23. **Folkman, J., Langer, R., Linhardt, R. J., Haudenschild, C., and Taylor, S.,** Angiogenesis inhibition and tumour regression caused by heparin or a heparin fragment in the presence of cortisone, *Science,* 221, 719, 1983.
24. **Shing, Y., Folkman, J., Sullivan, R., Butterfield, C., Murray, J., and Klagsburn, M.,** Heparin affinity: purification of tumour-derived capillary endothelial cell growth factor, *Science,* 223, 1296, 1984.
25. **Fenselau, A., Wald, S., and Mello, R. J.,** Tumour angiogenesis factor, *J. Biol. Chem.,* 256, 9605, 1982.
26. **Sherbet, G. V.,** *The Biology of Tumour Malignancy,* Academic Press, London, 1982.
27. **Rieber, M. and Rieber, M. S.,** Metastatic potential correlates with cell-surface protein alterations in B16 melanoma variants, *Nature (London),* 293, 74, 1981.
28. **Steineman, C., Fenner, M., Binz, H., and Parish, R. W.,** Invasive behaviour of mouse sarcoma cells is inhibited by blocking a 37,000 dalton plasma membrane glycoprotein with Fab fragments, *Proc. Natl. Acad. Sci. U.S.A.,* 81, 3747, 1984.
29. **Yogeeswaran, G. and Salk, P. L.,** Metastatic potential is positively correlated with cell surface sialylation of cultured murine tumour cell lines, *Science,* 212, 1514, 1981.
30. **Kalimi, G. H. and Sirsat, S. M.,** Phorbol ester tumour promoter affects the mouse epidermal gap junctions, *Cancer Lett.,* 22, 343, 1984.
31. **Fitzgerald, D. J. and Murray, A. W.,** Inhibition of intercellular communication by tumour promoting phorbol esters, *Cancer Res.,* 40, 2935, 1980.

32. **Yancy, S. B., Edens, J. E., Trosco, J. E., Change, C., and Revel, J. P.,** Decreased incidence of gap junctions between Chinese hamster V-79 cells upon exposure to the tumour promoter, TPA, *Exp. Cell Res.*, 139, 329, 1982.

33. **Tomasek, J. J., Hay, E. D. and Fujiwara, K.,** Collagen modulates cell shape and cytoskeleton of embryonic corneal fibroma fibroblasts. Distribution of actin, α-actinin and myosin, *Dev. Biol.*, 92, 107, 1982.

34. **Imhoff, B. A., Vollmers, H. P., Goodman, S. L., and Birchmeier, W.,** Cell-cell interaction and polarity of epithelial cells: specific perturbation using a monoclonal antibody, *Cell*, 35, 667, 1983.

35. **Talmadge, J. E. and Fidler, I. J.,** Cancer metastasis is selective or random depending on the parent tumour population, *Nature (London)*, 297, 593, 1982.

36. **Talmadge, J. E. and Fidler, I. J.,** Enhanced metastatic potential of tumour cells harvested from spontaneous metastases of heterogeneous murine tumours, *J. Natl. Cancer Inst.*, 69, 975, 1982.

37. **Chen, W. T., Olden, K., Bernard, B. A., and Chu, F. F.,** Expression of transformation associated protease(s) that degrade fibronectin at cell contact sites, *J. Cell Biol.*, 98, 1546, 1984.

38. **Humphries, M. J. and Ayad, S. R.,** Stimulation of DNA synthesis by cathepsin D digests of fibronectin, *Nature (London)*, 305, 811, 1983.

39. **Greenberg, G. and Hay, E. D.,** Epithelia suspended in collagen gels can lose polarity and express characteristics of migrating mesenchymal cells, *J. Cell Biol.*, 95, 333, 1982.

40. **Sander, E. J.,** Labelling membrane constituents during gastrulation, *J. Embryol. Exp. Morphol.*, 79, 113, 1984.

41. **Harrisson, F., Vanrolen, Ch., Foidart, J-M., and Vakaet, L.,** Expression of different regional pattern of fibronectin immunoreactivity during mesoblast formation in the chick blastoderm, *Dev. Biol.*, 101, 373, 1984.

42. **Schnebli, H. P. and Burger, M. M.,** Selective inhibition of growth of transformed cells by protease inhibitors, *Proc. Natl. Acad. Sci. U.S.A.*, 69, 3825, 1972.

43. **Unkeless, J. C., Dano, K., Kellerman, G., and Reich, E.,** Fibrinolysis associated with oncogenic transformation, *J. Biol. Chem.*, 249, 4295, 1974.

44. **Unkeless, J. C., Tobia, A., Oosowski, L., Quigley, J. P., Rifkin, D. B., and Reich, E.,** An enzymatic function associated with transformation of fibroblasts by oncogenic viruses, *J. Exp. Med.*, 137, 85, 1973.

45. **Quigley, J. P.,** Association of a protease (plasminogen activator) with specific membrane fraction isolated from transformed cells, *J. Cell Biol.*, 71, 472, 1976.

46. **Oosowski, L. and Reich, E.,** Changes in malignant phenotype of a human carcinoma conditioned by growth environment, *Cell*, 33, 323, 1983.

47. **Koono, M., Ushijima, J., and Hayashi, H.,** Studies on the mechanism of invasion in cancer. III. Purification of a neutral protease of rat ascites hepatoma cell associated with production of chemotactic factor for cancer cells, *Int. J. Cancer*, 13, 105, 1974.

48. **Yamanishi, Y., Maeyens, E., Dabbous, M. K., Ohyama, H., and Hashimoto, K.,** Collagenolytic activity in malignant melanoma: physicochemical studies, *Cancer Res.*, 33, 2507, 1973.

49. **Latner, A. L., Longstaff, E., and Pradhan, K.,** Inhibition of malignant cell invasion *in vitro* by a proteinase inhibitor, *Br. J. Cancer*, 27, 460, 1973.

50. **Giraldi, T., Nisi, C., and Sava, G.,** Lysosome enzyme inhibition and antimetastatic activity in the mouse, *Eur. J. Cancer*, 13, 1321, 1977.

51. **Coman, D. R.,** Decreased mutual adhesiveness, a property of cells from squamous carcinomas, *Cancer Res.*, 4, 625, 1944.

52. **Hunter, T.,** Phosphorylation — a new protein modification, *Trends Biochem. Sci.*, 7, 246, 1982.

53. **Schliwa, M., Nakamura, T., Porter, K. R., and Futeneuer, U.,** A tumour promotor induces rapid and coordinated reorganization of actin and vinculin in cultured cells, *J. Cell Biol.*, 99, 1045, 1984.

54. **Terranova, V. P., Walliams, J. E., Liotta, L. A., and Martin, G. R.,** Modulation of the metastatic activity of melanoma cells by laminin and fibronectin, *Science*, 226, 982, 1984.

55. **Poste, G. and Fidler, I. J.,** The pathogenesis of cancer metastasis, *Nature (London)*, 283, 139, 1980.

56. **Netland, P. A. and Zetter, B. R.,** Organ-specific adhesion of metastatic tumour cells *in vitro*, *Science*, 224, 1113, 1984.

EPILOGUE

In these volumes I have surveyed a fairly wide area of research, dwelling on different aspects of cellular attributes and their relevance to providing an understanding of animal development. The choice of examples used was aimed at demonstrating that a rational explanation of virtually all developmental changes can be obtained through a fuller understanding of the role of the cell surface. Recent research has clearly indicated that the cell surface is the site of regulation of cell adhesive and motile behavior.

Besides discussing established theories, I have referred to many recent findings which, though not necessarily conclusive, are important in identifying future lines of investigation. The nature of molecular interactions will no doubt be the area of thrust in the future. We talk of specificity of interactions between molecules in practically every context. The phrase is a convenient cover for our ignorance of a very vital aspect of the explanations that we are seeking. There is, however, no doubt that developmental biologists are fully aware of this situation. It is clear that more physicochemical information is needed about the interactions among diverse macromolecules such as those constituting the extracellular matrix and the plasma membrane. This is an immense task. However, many recently developed biochemical and biophysical techniques will yield more and more information. Binding specificities are, in a number of cases, dependent on the structural variety offered by the oligosaccharides at the cell surface. Newer tools such as monoclonal antibodies and lectins with high binding affinity for specific oligosaccharides can be used for probing the membrane surface involved in the cell recognition process.

Studies of cellular behavior, and its control by the components of the extracellular matrix, hold out considerable promise of progress. Artificial macromolecular assemblies have already been tested to study the behavior of diverse cell types. However, there is as yet no method of constructing a three-dimensional matrix in which the component macromolecules can be manipulated experimentally so as to induce truly morphogenetic behavior of cells. Manipulating the behavior of cells in sorting out aggregates is an attractive approach which will no doubt be pursued in the future.

In concluding this book, I must mention another point. Life on Earth has existed with continuity accompanied by constant change. Biologists have greatly benefited from the ideas of evolution. In seeking a rational explanation for developmental phenomena, one should never lose sight of the general guiding principle that life has had continuity ever since it came into existence. We can profit immensely by examining a wide variety of living forms in arriving at generalizations that will elucidate embryonic development.

INDEX